TO THE FUTURE OF THE PEOPLE OF THE UNITED STATES OF AMERICA AND THE WORLD IN THE NEW WORLD OF RILEY MILLER'S PHYSICS—

Riley Miller

"Riley's Natural Naturalization"

Authored by the great; Riley Parker Miller √XO

> Copyright © Riley Miller 2012
> All Rights Reserved
> **ISBN-10:**
> 1479395447
> **ISBN-13:**
> 978-1479395446

3

"The Naturalization Theory's Outlook"

Riley's Natural Naturalization

Warning- This is the book that happens to be, on salvation that is life, on life. This contains interests that are about the areas of philosophy, which have science and religion in them, with also church. This is hell, and a heaven, has a life. What this pertains, to, is about hell and heaven. This is about, the life of the church, and about the real life, of body. The healthy churches contain government and the body of the people that are about irony, and satire, and the life of humor for others and people. These are the person, who is about the life, which is about the reality

living, which is about the busy life that is about people. This is not, merely a wrongfully accused and anonymous and then chosen life manual instruction book. This is the chosen Biblical Instruction Kit, for the living and the dead that I hope you can enjoy, and contextually read. This is about life. This is about science, and math, and about how the ways, of intelligence, is about intellectual and honest, ways that are about Heaven. This is about church, and Christ, and the life of Heaven. It is about honest exploration, and about the interests, that are seemingly, in the body of Christ. This is of, and about, and containing the conflicts, of the areas of the real level from interests, between sciences and the real chosen church. This is about conflict, and the areas of interest, between the known and unknown, that are about the living and the life. This is about, the living and life, of the chosen people, who are about and for, that are about and from, the life of God, and about the real chosen life, of life. Life, is about the contextual real life, that is about and of, the real life, that are a chosen, and a normal, way of living life, to the fullest, and utmost extreme. Life is about Jesus Christ that is about Jesus Christ. Life is all for the life, that are the chosen, and the way, that is for God. God is Jesus Christ. Natural ways, that are symbolic, and meaning a lot, that can and will, have the surface of heaven, but is in between, heaven and hell. These are ironic, and symbolic, through the ways of Heaven, known as the Garden of Eden's, passageway, that are through the eyes of Jesus.

6

By: Riley Parker Miller

Riley's Natural Naturalization

Dedications:

"This book is dedicated in to what is in between sciences, and the math, and what makes the real world seem much better, to me, as the all abiding, title, of the real President King, and the real "King of All". This is through all the works, and the faiths, of all the individuals' inferences of the abroad world, that are about the real world of Himself, the real Jesus Christ. These are of, and able to be within through the real world, that is about churches, and the progressions, of sciences and mathematics, in which is all about Heaven, and from, all the real deals, of the church body and the life, and helpful living.

"God is about math and science also, in a manner of speaking, but He is about, the true nature, of faith and works, that are about confliction and interests, of the words that bring the ways of the world to science's doorstep, in faith."

I like the church and the science behind school. This is why this was written.

The Introduction:

How Naturalization Was Born-

I have always thought in terms of something and I did not know what it was. This is an exploration. This is of the

unknown. This is how we think, and all act, and all do. This is the basic.

The thought behind the thinking that is behind the thinking is about the way deformation is about the world. This is about the way the world is about the thinking that is about the world. This is due to synergetic thinking. This is what the world is about that is about the way things are coordinated with us.

The thinking that is about the way thinking is about the way thinking is about the way thinking is about us. We are each individually related with synergetic thinking. This is about the way the synergetic thinking is about the individual. This is what the individual is about and what goes into thinking. This is about the way thinking is about thinking. This is absolute and smart.

The way thinking is about the absolute is about the 1. Standards and 2. Recollections and 3. Duty of thinking. This is what the thinking is about that is about naturalization. This is a book to sum up all my books, as an introduction. The way that "smart" is about the way we look, is about principles, which dictate this with a social standard and a recollection of past events. No one could be smart about the creation of the world, yet could be "absolute" about this, in a particular way.

The way naturalization works is through synergetic thinking. This is about absolutism and about principles. Anyone who does not know naturalization is "smart" in a different way. This is about an example I will set for society. These books, have been read, and recorded in the "Book of Life". This is what Jesus Christ has after we are judged. This is a whole outline of my life. I am not worried. This

is about absolutism. The creation of the world is through this.

The standards are about the one individual notion, of independence. This is of my flavor of essence and substance. This is about writing, and the unknown, and the flavor of this. I hope my afterlife is the best.

The way we judge and are judged in the Book of Life is about the substance of life. This is Jesus Christ in whole. The way we are judge and jury of our own lives, probably is not acceptable. This means that Jesus Christ has dictated our whole lives.

This is the Jesus syndrome. He is God.

If there were not Jesus, there would be no God, and there would be no judgment. This is not what I think about naturalization. I think this is a smart, absolute, theory. There is no judge to it.

This is because I am the one who is judged. This means that the judge sits rightly on the right throne of God. He created the universe. This is the details about all that followed.

This is about the diction and circumstance. This is about the dictate and circumstance. This is about the way the world is about the direction and circumstance. This is about the whole intellect and substance of the matter. This is about the way that the entire creation of the world, is about the small and unusual. I was not inspired to write naturalization, but I was a writer.

This is of the whole of naturalization. It is not a judge's theory. This is about parts of the world and how they naturally interact with the substance of living. These are the stories of existence. These are about the smart and

unusual ways of naturalization. These are the parts of existence, as we see them. They are not about the elements of fire, water, and earth. They are about the elements of existence as we have seen, and known through the creation and after it. This is the theory of knowledge and how they work. They are 1. Perception, 2. Naturalization and 3. The outlook of creation.

These are from the synergy of thinking.

This is for anyone and all of us. This is how the creation of the world got started. Yet, this is known through after the fall of Adam, because of the knowledge.

My theory is that God created the universe. After this, we fell, and fell into knowledge. So, we thought that we lived after the fall with this knowledge. This is what God had.

These are the dictations and substances that we all acquire and know about. These are for and about the creation of the world, and how it is done. It is done with a Bible. This is why Jesus called it the "Word of God". This was always in the first beginning.

My theory of naturalization is about what like to call, what Jesus had after He came.

This was also a way to create the universe. This heralds too many questions.

I do not have a rock hard faith in Jesus, but I do have knowledge of what happened. This is the Holy Scripture. This is about my faith with naturalization. This is about how it is hard to knowledge, within, and without, have within. This is within the knowledge of the creation. All my point is that I have not witnessed it.

This is how the Holy Scripture is about relational philosophy, and about how the philosophy is about the universe. This is from and through the essence and circumstance of knowledge and knowing. These are what the essences and qualities are about that are about the way they are from, with the creation is about. This is Jesus Christ. This is a whole theory about the creation of the universe. This is how the creation of the universe is about knowledge.

This is also a relational theory, because of knowledge. So, naturalization is about a theory that is about the way the world is about perception. We are this in our beings. This is what knowledge came from and where knowledge exists.

The whole world's illusory perspective is that the world was created by Jesus Christ. This is obvious fact that is not scientific, but spiritual. This is about the ways that the control of the universe, is controlled by evil, and evil, impulses that are about the absolutism, of natural perspectives, and how they originated.

This is not, about, and of, the natural inclination to sin. This is not about naturalization. Naturalization, is a study to find out the true meaning of, creationism, and with the absolute and smart ways it is about and exists in a created outcome of the universe. This is about the ways that the empirical truths is about the essence and conflict of knowledge having always been around and have been for. This is always Jesus Christ. Yet, how did Jesus know about the conflict of interest in the world, with illusive qualities, that cannot be answered? They are answered by the knowing of the naturalization. This is a synopsis of existence.

This is about the ways that existence and qualities of existence are about and from, being created. This is about the knowledge that we fell with. I consider this theory as implausible with Jesus Christ. Yet, even though He created the universe, He came after the universe's creation. He was also man, and this is why He created it. This is also from His "man" perspective. Yet, every being sees naturalization, as someone who can see.

He is Jesus Christ. Yet, He did create the universe with a different perspective. He was the pioneer.

This is about how naturalization is about the direct and opposite challenge, of creation, according only to the evil. Functionality, in the ways of man, and his environment are about the direct opposite. This is about survival. The "survival" is about the way that man answered. He was challenged with death, and labor, and perils of man, and how he functioned. This was the 1. Effects, of knowledge, and not the 2. Outcome of intellectual intelligence. The naturalization of man was about the outcome. This was also God.

The theory of the outcome of knowledge, is about the complete and complex realities of God, that is about the way the God of the world, is about the natural ways man acts. This is from the product of naturalization, and the way we have perception. This is about the way that we act. This is from an acting agent. This is of temptation and how we fell. This is not philosophy or literature, but about the ways we act, and are in and for control.

The acting agent of desire was about the way we fell with knowledge. This is about temptation that is about the complex realities and how we are from them. These agents of acting are about the product, and naturalization,

of the outcome. This was fully because of temptation. This is how acting agents help us, and become with us. This is about the ways that the old, and new, are affecting us, and how we learn from this.

This is about the way the locale is about the reaction that we are from, in the circumstance. These I called in college, circumstantial happenings. These are about acting, and the rationale behind this.

These are the ways that we see through the essence and substance that is about the complex and complete ways that desire is challenged, and made new. These are through the perception of God.

God is the first ingredient in naturalization. This is about the ways that naturalization is a part of society. He is a part of the nature and man that is about the fall.

There are so many different ingredients that are in naturalization, that it doesn't make sense. This is from a universal perspective. This means that man is the acting agent of himself, and desire and temptation is how we fell. This is from the Garden of Eden.

This is why I call it naturalization. Due to the fact that naturalization exists and is existing, is a challenge, behind which the naturalization is about success. This is just from one ingredient of God. He is about the ways that Jesus Christ, is about Himself, which taught man this.

What man is about is about the natural ingredients of his perception. This is the philosophy "senses". These are what we see with, and experience, with what dictates us. This is about a math and philosophy theory that is born. This is about the way the world works. This is through this theory.

The philosophy behind this theory is about and due to a philosophy. This is about the way the world works. This is through the book theory that is about the ways that man excels, through the outcome of the theory, from the senses, and about and through the dictation of the senses, that is through math and philosophy, which makes sense about this. This is the outcome of a conclusion. This is what the theory of the outcome can become. This is from and about the dictate and ruling of the senses that are about the philosophy that is about the dictation that are from naturalization. This is about the way the world is about the sense of being, that is about the rules that is about the philosophy. The philosophy of math and science is about physics. The outcome is about the way the world rules over the ways the rules are about the science with the philosophy that is about an art, and a science.

1. Philosophy
2. Art
3. Math
4. Science
5. Written words.

These are the whole of deformation, and how it was created as a new language. This is how it is explained through physics. This is about the way naturalization was born. This is about the ways that naturalization is about the ways that science, is about the ways that naturalization is about the birthplace and origins of deformation. This is about a science and math with a written language. This is from the unexplained. This is how it became real. This is through progress and appeal to the real humanity, of the person.

Naturalization in a Religious Context-

This is about the way that naturalization is about the freedom of the choice, and how people, are chosen to be part of this. This is about a theory of the world and how it has to do with creation and origins that are not ancient but learned in the sense that it was before creation. This is about the creation that is about the origins that is about the philosophical context. This is about the way that philosophical context is related to religion. This is about God and Jesus and not about science and math. These subjects are not the real way that religion is created, and example wise "known". This is through examples of philosophy and not parables. The religious context of man and woman is the Bible. All people know this. We are all related to Jesus in one way shape or form. The fact is that we all know God. He is the one who thinks for all of us. He is the one who created the world. If this makes sense, then naturalization makes sense to all. He is the one who is 1. One, 2. Two and 3. Three. This is about God in three persons, blessed Trinity. This is about how we are about the direct and in front of, thinking. This is through three persons. These are the "persons" of the Trinity of Jesus Christ, and God, and the Holy Spirit. These are about the three persons that are about the blessed Trinity, of God. He is the one who makes us all think. So, three things, makes sense, especially because God is about this. He is the designer of the religion, of God. This is how the Spirit of Truth makes it happen. This is about the way the world is about the good, and the direct opposite known as evil. This is about how the cowardly and ungodly are not part of the Trinity. So, they do not think in terms of God.

They think in terms of collaboration with other people. These people are the scum, of the earth. They do not know religion or do not know coordination, with Jesus Christ, who is God, of all three things, that are Christ's body. These three things are in coordination, with other things, known as the Trinity of church, the Trinity of Heaven, and the Trinity of earth. These three institutions are fact. These are about Jesus Christ. These "make up" the Trinity of the whole world, that was the Alpha, the I Am, and the Omega. This was before Jesus' time. This was naturalization that is making you realize this. This is because naturalization is a theory, and is not a pure intellectual premise, guarded by Truth. This is the Holy Bible. This is what the Holy Bible seems to know. These are that Jesus Christ is the "Word of God". He is the one who knows people, and how they are. They are God.

Naturalization in Terms of a Theory-

This naturalization thing is guarded by truth. This is because I am writing this after telling you that it is not part of truth, yet part of a theory. If I am speaking in terms of a contextualization theory, then the context here is about the naturalization in terms, of a theory, and not the truth. But, if I say it is the truth, when directing the thought toward nature, man, mankind, God, environment, naturalization, then I am telling you that this is true. In naturalization, we should beware exactly where the truth is when we are seeing it. This is about the truth of relativity and how this is also, contextualization, which is commonly referred to as the, "absolute truth", which in fact, really does not really exist, in thought and in thinking, for better

or for worse. This is merely, philosophical thinking, that is guarded by Jesus Christ. This is ho w all absolute theories on truth, are about. They end up centering, on the whole of the Bible, and how it relates to everything, yet, this theory is physics type. So, yet we know it, it is not unusual. It means that this is what we think, when we think it, and how we think it. This is through this same "as is" kind of thought. This is naturalization in terms of a theory. This is meaning that the one theory, of thought, is related here, in terms of a theory that is about the context. The context of absolutism is a theory that is also, about naturalization that is also about the unexplained, which is thought. Yet, naturalization on the other hand, is related to thought, which is about the context. This is the great divide that is thought. Deformation is about a kind of, thinking that is about the thought that is about the thinking that is about the true. This is because it is believed in by the absolute necessary. The factor of naturalization is about the true, and fact, and for sure, ways about which are about the ways, that are about the God. These are about the fact, and the real, and the true. The division line between naturalization, and the ways that naturalization are about, is relative. This means that it is a true theory. This is about how the theory of relativity is about the context of making, realizing, and generalizing, what the standards of society, are. They are about the complex and complete, and obvious, ways that we learn and know about the theory of philosophy that is for and to knowledge, of the fact. These ways that we see knowledge are about the ways that knowledge is about naturalization that is about deformation. In conclusion to the introduction, the ways of knowing, are about naturalization.

The theory of naturalization is about the ways that naturalization works, which creates the language of deformation. This is about the dynamics of the theory that is about the realism that is about the physics. These are the laws that apply to naturalization, that are always in the realm of knowledge. All of these laws, are about, the knowledge that are about the ways that knowledge work, that are about the theory of knowing, that is about the dynamic ways that physics, is about the unexplained. Then, it afterwards, is about the explained ways of thinking. This is in terms, of absolute, real, and the outcome, and scientific analysis, and the awkward expression of man, known as God. This is just a perspective of man. In math, it is about the progress, and gain, of fully real and known understandings, of the worldview of knowledge, and how the senses know all, about physics. In the realm, of the Biblical, there is knowledge, of God, through all of these. In terms, of atheists, this is about the touch of glory, and how the world, abroad affects this view and something from this, known as Satan, or a false thing, or a non-religious perspective, that is also somehow justified by God, known as Jesus Christ, in the ending, of the Book of Life. This is strictly from a religious perspective. Yet, I have known this apparatus and functionality of man, and his exploration, of known business, and conquest, and ventures, of money. This is his whole life. Yet, after he fell, there must have been a theory that is about his way of seeing, with knowledge and without knowledge, from the fallen man, and his existence. This is about the pure, and simple, ways of knowing. These are about God and man and how He communicates, well, through the dictate, and know, and realize, of the expression, "naturalization". Naturalization is defined as the conquest of something that is greater than it. This is deformation. This is about

nature, man, environment, God, mankind, naturalization, and the facets of pure understanding. These are a theory of pure understanding, instead of knowledge. The fact is that man is about the complex and complete knowledge of himself. This is the theoretical philosophy that I cannot ever get past. The complex and complete world of philosophy, instead, with this theory is frankly naturalization. This is about the years that naturalization, has gone through, and known, possibly ever since the dawn of the fall of man. This is a first theory and a last theory of knowing, and knowledge of difference. This is how we see other men. This is a theory that is involving the senses that are somehow related to these. Man would be the sense of sight, environment, the sense of intellect, nature, the sense of knowing, and the sense of mankind would be the grasp of intelligence with knowledge, and the God would be how we see all things subjectively. Yet, with naturalization, like the art form of deformation teaches, is about the words and expression, of one and another, that one chooses, that believes or doesn't believe in something or another, with existence. The expression of deformation, is about creating your own perspective, and seeing what you believe in, about the close world, of deformation. These are cleverly, instigated, through the known, and the knowing, and the knowledge. A line in deformation could be, cleverly stated, as, "something that is unusual to comprehend". This is objectivism that is already explained in naturalization that is about the absolute God.

He is known as the creator of the world. All else falls behind Him. Yet, we pay little value to this notion, in the modern day of independence, with a country and a world, to know. This is the introduction to "naturalization" or

the "cultivation of culture in another person's country with laws that make one person, "naturalized", as a citizen or resident, of the new country, with the laws of the country applying to them now. This is about how 1. Something that is unusual to comprehend is 2. The cultivation of culture in another person's country with laws, that make one person, "naturalized", as a citizen or resident, of the new country, with the laws of the country applying to the m now". These two points go together, with the spiel. This is about naturalization. "Welcome to a world of deformation". I am your introductory speaker. Or, you're evil. This is about the way the world, sees naturalization, and about the ways that naturalization, is about naturalization, is about the created and outcome, of itself. This is about the complex and unknown realities of Jesus Christ, and the God of the universe. This is something even Stephen Hawking could easily master, nonetheless, also invent. This is a real theory that is about the complex and complete unknown, that will introduce you into the world of the world. This is your ticket to the universe. Thanks. Riley Parker Miller.

TITLES AND POSITIONS-
The world is a vast and complete place, where you can 1. Raise your kids, 2. Raise your life, and 3. Get used to your environment. Yet, no one explains world citizenship like this physics theory, naturalization, does relate and eventually goes for. This is about fact and thinking, that is about the complex world that is about the completely unknown realities that are about the world, that are about the complex world, that are about the complete and unknown world of fact. This is about the scientific that is about the usual that is about the complete known world, of fun. This is also about naturalization, and its politics,

and phrases, and known worlds. These are of the world of creation. The theory of naturalization is believed in because of the language of deformation. This is what primarily my books are written in. They are art and math and words that create and condemn the world. These are about the complete and obnoxious world of books still that create one another, and react with one another, in a good way. This is toward the notion of freedom, and the notion of slavery. This is about the survival of the fittest, and the naturalization of the universe. Related to these four classifications, there is a reality. This is of the freedom, the notion of slavery, and the survival of the fittest, and the naturalization of the universe, and the responsive to these four categories, of world knowledge. These are the reasons why God created the world. This is if deformation is true. Firstly, the survival of the fittest is false, and does not explain anything with truth. This is the reality of the underworld. This is why knowledge cannot get past anything. Second, naturalization is a universe theory, because it is relative with anything, due to the theory. This involves, as a warning, slavery or freedom. Third, the reason why we are in a context is because of this. Or, we would be in the Garden of Eden still. Yet, we are slaves to death, work, childbirth, sorrow, commitment to marriage, and the snake's plagues that are the worst. I am a person in my life that is about the client that is of marriage that is of death that is of labor that is of commitment that is also of knowledge. This is about the love, of money, and primary care, and physical obligation, and laws of the world that are about the client and apprentice. If I am a weasel, or a snake, or a God or a man, or a woman or a lover, I am still not a man, known as God. This is the boldest statement ever. Let me explain.

I once, was born, inside another. This is fiction. Then I became illegal. This is the substance of the few. To take my old writing, before deformation, that is about nothing. This is because I did not know science, and math, naturally. This is because I had no art form. Yet, I called myself, fiction as something illegal, defined as a "good one" for the world, of the "illusion" and the "make believe" in books. Now, I am satisfied knowing all thought is true, and because of this, I created a word bank known as deformation. The only real explanation of how the world was formed is through this theory, as of when I am writing this. This is about the client known as deformation. I am the apprentice. I am also a person, who is well united with the truth. Yet, I do not know what a gentle man knows. This is his own burden. This is to go well, with others, and behave. Yet, I choose to go against the grain, and progress. This is to resist temptation known as the sin, and believe. This is in a wonderful maker. If we did not have this, we would not be able to think. This is how the Serpent tricked the world. He made an image of himself, above God. This is why he "crawls on his belly". He is the deceiver known as "Satan". This is about the won, and the loss, and the winning, and the losing. The slogan of deformation is to "know thyself" and to "think of all and all of it". This is about all things that are about creation, and how we only have knowledge, after we fell. This was the knowledge of God. This is why we are not there anymore, unless Jesus Christ lets us. This is about the way the philosophy of God, is an uncommon and strange happening that is about circumstantial happenings. These are the happenings of truth, and compromise, and knowing. These are the contains of knowledge, that are about the diction, of the world, and how they are about the world. This is about the way that the world is about the context

that we major in with excellence. This is from a construction of society known as "school". This is about the way that we are condemned, or successful, and how we know this, is through the direct, and oppositional, ways of man, and how man is responsible for knowing. He knows through the direction of knowledge toward the future, past, and present. This is also with naturalization. The whole world can be a product of your own knowledge. This is about the direction we talk in and walk with. This is about the way the world turns, and without knowledge, there is no major thing, that is about knowing, that is also, to know about, the real existence that we have, and how we know of it. This is through the understanding of God, with the knowledge of Him that is about the direction of knowing, that is about knowledge. This is about the closure and commitment with God that is about knowledge. This is always men. The knowledge of God is like a snake. It is commitment oriented too. This is about the complex and complete ways that knowledge is about the knowing, and showing, of the world's commitment. This is to the God who created us. This is about the way that the world moves and rotates. This is all purely the fact that God exists. This is about nature, man, God, mankind, naturalization, and environment. This is about the way that God is about the nature of man. This is about the way the God of the world does not make mistakes. He is perfect and we are not. This is about the way the God of the world is about the perfection that is about man and God. He is about the conflict of interest that is about man. This is about the way God is about the way man is about the way God is about the way man is about man that is about explanation, in which is about the way we were created. We fell because of lust after God's knowledge. But, this was considered desire because of the way it was about tempt-

ing. This is about the nature of God that is about the nature of man. He is about the confliction of interest in the fall, of man, because of sin nature. Sin nature is the ambivalent explanation to the creation of the world. He is passive and resistant in his own thought of tomorrow, the world, and the fact and fiction of the universe, being created this way. The way that man is about God is about the necessity of God, and about the fall that man is about is about the natural instincts and reflections of the sin, that took us there. This was always about temptation. This is about the way that God made Adam perfect, and since then, he did not know what knowledge of good and evil was, and ever could become. This is about the way that knowledge is about the way that knowledge is about the fall of Eve first, then Adam, based on the Serpent's temptation. So, realistically the snake doesn't have knowledge, but just tempted Adam. From these perspectives, we can see the fall of man. This is comparative to God's perspective. God sees the world as 1. Good and 2. Evil and 3. His. The next is Jesus who sees the world as, 1. Holy, 2. Good and 3. Righteous. And the man who is also a Christian sees the world as 1. A combination of good and evil, 2. Holy and Damned, and 3. His and also Christ's. This is what makes up God. This is about God. This is of the existences of God. And this is because of the existences of God, and Jesus Christ, and man, inside of naturalization. These three key examples are the compromising of the world, as in the view of God. This is what explains the theory of nature, man, mankind, God, environment, and naturalization, and as well as the theory of what makes this up, and on, and about. God sees man as good, because of Him. Yet the theory of naturalization is about the theory of the way coordinates create and destroy. These are about the creation and destroying of one another. This is about the way

the world, does things, that is about the way the world, is about the strength and ability of Godliness, and good, and evil, and combinations of this. This is because naturalization is a theory. This is also containing good and evil, knowing, that is personal, and perspective generated and known. This is the theory of how good generates itself, much like environment would explain, due to knowledge, as well as man, mankind, nature, God, naturalization, and the causes of this. This is about the way that angels are not. This is about the way God is only certain of this theory. This is about the way subjectivism rules the world. This is through the math, and science, of the art form of knowledge, that is thought and thinking. This is about a world that is through the process, of math and science, and how these are not just in thinking. These are part of the explanation of reality.

Thinking and Deformation-

(An Allegory)

The thinking of the process of deformation comes from the Bible. This is a tangent, realist, expression, of love and good. This is Jesus Christ, who knows all of the future, with knowledge and God, and also Jesus Christ Himself. The theory of the nature of man is also included in this format. This explains that the way we see and view the world, is through math of science and art and believing in this theory. This is about the way the world is about the complex and complete and combination realities of good and evil, which is the primary basis of a theory of knowledge. This is about a combination, of good, and evil, that is about prime and generational thinking, that is about all that it is concerned with. This is concerned with the knowing and knowledge of the fact that people are

alive and well. This is concerned with the fact that nature is a basis of human knowledge. This is a fact that, "We" created the "Ourselves" in the beginning. This is the explanation of God. I have always struggled with this explanation that is about the conflict of the world, and how we must have learned, about and with us. This is the explanation that how Jesus created the world. And, also God must have also created the world. Yet, we do not have origins of this creation, in fact, and the world, and in religion. This is other than God's voice. The method of the world is through and about, the theory of what created it. This is not a lie or a cheat, but only a theory that the world was created in this context. This is about the context analysis. This is of God, and Jesus Christ, and man. This is about how knowledge could have been there before time, because of the way that this mathematics is concerning, is about the conflicting interest of everybody. This is a theory that created the universe. This is a theory that is a lie. This is a theory that is about concern with the world, and how we created this. This is about the voice and how it must have been created as well, through God. The fact that God, created the universe is about the way that the mind is about the material energy and that there is no witness to the creation of the world. Therefore after the math is complete, there is a void in the beginning, and the voice could have come from Jesus Christ. This was in the beginning, with the Word of God. This is complete and total harmony about these issues of God, and about how the world is complete with Him. He is a total mystery but not to the "voice" but there was a formless void before the voice spoke, and also, we do not know how the world was created personally, or we are a lunatic thinking this. The Bible is the divine source of inspiration and contemplation for anyone and everyone. It does explain things. It

does judge and see things. And, with this in mind, it does create and help explore things that are of interest to us all. These are the things that do matter to us, and do rely on trust, and do see the way things are, and are from. The ways that the new and old have always come together are solely about the essences, of the creation of God. This is about how the essential attitudes of the masses are about the control and conflict of the reason that we are about the ways that are about the God of the world and about His conflict. This is about the world. These are the evils of the world that is about creation. These are about the evils of the world that are about false theories. These are about the conflict of judgment and knowing who is good, and who is stale. This is about lukewarm, and hot, and how these are explained. These are to the God. The factual existence of man is about the luring of God into the open, which examines Him, and leaves Him open to anyone who would like to see this phenomenon. This is about the factual evidence that is truly His. This is about the factual examining of what man is, and about how man has come to the conclusion that He is God, especially after Jesus Christ came, to the beginning. This is about factual existences that are about the hope that God, has in it. This is the world's view of examining the world, and the ways that the world has been seen, and about the arrangement of the challenge that is about the open. This is what the "open" is truly containing and concerning. This is the problem that exists in the world that is about the examining of the truly great emphasis of man, in God, in the beginning that is about the "open" issue, as I have hereto explained and seen which is about God. This is what God, has intended. This is to have this problem, open, and concerned with. But, this is a "gay science" according to great people. This is like Nietzsche depicted in his book, "The

Gay Science". In my opinion, he left God to be examined, and depicted in a translucent cross. Yet, he was not sure that God was "open" and what God was about was about the being, of creation, as His, after He spoke. According to skeptics, and according to me, there is a gay science about the universe. This is that when we are on the planet earth, there is a spectacle to be skeptical, about. This is about the ways that scientific meaning, is about the meaning, of mankind, in the natural ways that he sees things. This is actually the most brilliant theory that has ever existed. This is about the ways that channels of energy, are about the ways that mankind, has opened up, the universe, and has left, no cloud open, and no mountain unturned. This is with his knowledge. God's knowledge about man is that He is open, and turned into stone. This is about the ungodly and how the government and about how the people, are existing, in existence, with the worthwhile Godliness of the Godly. This is about the meaning of naturalization. This is about the relationship that the ungodly, has, with the open, and challenging, ways that man and mankind, has examined, and cross examined the God, of the universe, and how He has no flaws, except with the ungodly. This case, will never be examined, and concluded, that the open case is about the exceptional qualities that are about the ways that are about the exceptional ways that are about the ways that the ungodly, are never about the cause, that is about the great, and strong, ways that are about the conclusion of man, being what man is about. This is about sin nature. This is about the ways that the man, and the ungodly, do not relate, and do not have any problem, with each other. This is about the way that people relate, that is about the created outcome, from God, and how we do not see this, ever, in any existence, or in any situation, known to man. This

ungodly, is a choice that can make it with the world, that is about the problem, that is about what the problem, is about, that is about the ways that the problem, is about the quality, and complex, ways that are about black and white. There is no human race, according to the "gay science", because it is the "ungodly". This is what Nietzsche, the great writer, is about, and what he concerns himself, with, is the quality of man. This is in the ungodly position that he comes with. This is about the way they contain, quality, and compromise, are about the quality, of the use, of God, with the world, and how he sees man. This is through the eyes of whatever he sees, to be quality, and compromise, and contain which is about the philosophy. The "gay science" of Nietzsche, is about the ungodly and its creation of the outcome, that is with and in the world, with the world view, of God, with the quality and containment, of quality, and compromise, that is about the world, and about what, the world is about is about George W. Bush. He is about the compromise, and quality, and containment that I have always liked and appreciated, in the world, of the illusion, and the weakness of the world, and its compromise, and how its compromise, that is about the ways that the world, is. This is about the sane.

This is about the sane. The sane is about the creation, of the intellect. This is about the sane.

A. Corruption and the Doctors Concerning Naturalization-

These facts, about and from, the concerning, about the issues, are about the problematic, issues about doctors, and about how the issues to explain doctors, is irrelevant. The issue, concerning doctors, is about the concern, about how doctors are really there? This is

about the concern, of issues, regarding hypocrites, and their targets, of the people. Do they really see, these people, or are they just concerning themselves, with other people's problems? The doctors are about, concern, compromise, and qualities, of lying about diagnoses. They are the intellectuals. This is why I like them.

The qualities of naturalization are not for the father, and mother, to talk about, and make sense of. This is especially, in the context of doctors.

B. The All Encompassing Thought of Doctors-

Doctors have always had and been diagnosed with thought and thinking, to the context, of thought control, and thought manipulation. This is about the complex and situational, disease, of doctors, and how patients, have helped the ways that life is about the context, which is about the contextual disorder, that is about and according, to the context. This is of the grand design of the universe, and how order, and disorder work. This is about the context, of manipulation, and how the manipulation, of thought, and thinking, is about the complex, and situational thought, that is about disorder. This is about the manipulation of thought. This is exactly what is about, the ways that I think, and how I think, this is possible, is through, the process of thought, in doctors, and lawyers.

They are the authority figures on everything.

THE DOCTORS WHO ARE CORRUPT, WITH LAWYERS-

Some people get committed, and do not succeed. This is due to corruption. I do not like doctors and lawyers who are corrupt. We think that the issues, to succeed, and to detect, are corruption and smart and sane ways of thinking are better than corruption. These are the successful; ways of thinking that are about the context, and contextualization, that are about context, and design. These are about the concern, and compromise, of great, and small, relations that are about the complex.

Corruption, Concern, and Authority-

These are not authority figures. These are about the figures of the authority that are about normal, everyday practices, that are about context, and about the compromises that are about the conclusive evidence, that all things are either, fake or real. These are about doctor's opinions. Also, this is about corruption. And, in addition to this, there is about the unreal. This is a cross between fake and real, because the fake and the real are coordinates in the logic of mine, according to doctors. These are about the unreal. These are about the unreal or the imagination that is about faking it. The fake, the real, and the unreal, are coordinates of the axis of the coordinates of the naturalization.

These are where the authority on the basis of thought exists. Either a God, created us, or a scientist's theory, or a God whose love is in something else, or a scientist. The doctor, is a pretend God, that is about the conclusion that is about the authority, that is about the evidence, that is about the conclusive, evidence, that the authority of make

believe is the doctor. Science explains health, and it explains creation, of similar things. Yet, science does not explain the creation of the universe.

This is the allegory of the doctor.

The Allegory of the Patient-

The God of the universe is supposedly, about the issues that are about the context, that are about the conclusion that are about the knowledge that we all have. This is about the context that is about the conclusion that is about the real, and the fake, and the unreal. This is about the context, of the thought, with function, with authority, with the exception, of the unreal. This is about the patient.

The thought functioning of the theory of knowledge, is about the functioning, that is about the conclusion that is about the context, based on evidence. This is of the outcome. The outcome of the theory of evolution is about how the conclusion that is about the context is about the behavior that is about the condition that is about the context that is about our temperament that we are in. This is about how music inflames temperament. The music of all of us is about the movement from, one cord, to another. In perfect contrast, the music conduction is about the explanation. The movement of music is conducted, by the experiment that is about the context that we all see the construction, of the music, in. This is only with our ears, eyes, and all other senses. This is about the control of the movement that is about the necessity that is about the senses, that is about our control. This is not over the music. The movement of the music is about the way we accept it. Yet, this is irrelevant to the conduction, and pro-

cedure. This is about the way the music is about notes. The conduction, of the music, is based on the surrounding thought that is about the relation to the constructed seeing, hearing, and believing, in the music. Music is the great communicator, and how we see and perceive it is only up to us. Yet, it is a collective outcome of witnessing, like law, and creating it, on the other side, like a doctor would. This is about the collective outcome of the outcome that is about the outcome. In words of deformation, the outcome is about the complete, and total, harmony, of the creation. The creation of the music is about the created, outcome, of music, and how this created, outcome, is about the indifference or acceptance, of the created manuscript. This is also how a book is written. This is about the science, and math, and philosophy, of what is and what can be. This is how we can become with music, the great communicator. It is witnessed in the eyes of the beholder, which means that the music is about the ways that the control of the world, is about the science, and math, and philosophy is constructed. This is with ease, completion, and scientific outcome of the outcome. This is a language of philosophy. This is about the language of philosophy. Yet, this is in the outcome of the language. This is about how we see the results. The languages of philosophy are about the essence and substance. This is about the language of philosophy that is about the calling. This is about the calling. This is similar to a philosophy and a theology. This is all meaning that the world, with the definition of deformation with its impulsive relationship with the world itself, is really a theory combined with philosophy and theology. This is because I have written a book. This is in the real Bible. Yet, it is not yet. This is similar to a philosophy of deformation. This is the theology lying. This is about the creation of the world.

Riley's Natural Naturalization

Lies and Deformation Do Not Exist-

The language of deformation is about the context and the language of the outcome. This is about the ways that the deformation of the world, is about the context and opinion. This is through the outcome of the philosophy of the world. The ways languages are about, the context, is about the context that is about the contextualization.
This is what is in God. These are in God, and what is in God, is in the outcome of God. The equation and the theory explain this. This is from and about the context that is about the contextualization that is about the new and old, and how they are explained. These are the projects of the outcome of deformation. This is about deformation that is about the outcome that is about the theory that is about the outcome. This is about the ways that the outcome is about the 1. Philosophy, 2. The theology and 3. The Biblical Interpretation. These are about the challenge that is about the difficulty that is about the wonderful ways. This is about the ways that the Bible works. This is compared to science. In science, there is no God. This is until now. Yet, the science will keep on explaining and keep on knowing, what the world knows, in a different way. This is in a lie. Yet, the Bible is always true. All naturalization is, is an explanation. This is on the Bible. This is of what could have been, and will be, the naturalization of man, that is seen. This is through the theory of this. This is also through the given outcome of this venture. This is about the ways that the world is about the context that is about the reality. All of this points to God. This is why naturalization is not a lie. This is because God created the universe, for all of us. Yet, the Word is simple, and also did it.

This is the Word of God. The reality of the given outcome of the God, of the universe, is about the simplicity that is about the creation of the world. There is no hypocrisy in the universe, except for people trying to take credit for things that God did not do, but in His name. These are the only people who think differently than the truth. These are the people, who know about the world. Yet, they know about it with sin, without being "forgiven". This is about the ways that the sin, is about the sinning, that is about the nature, of sin. This is wholly naturalization and the nature of the world. This is about the world that is about the ways that the world is about the context, of the world, that is about the relationship with God. He is about the context of "becoming". This is the whole context, of what the deformation of the world is from. This is that man does not evolve, but instead, becomes what is in the Bible way of seeing and knowing.

DEFORMATION IS A TRUTH THEORY-

The language of naturalization is about the creation of the world. This is one way that the world could be explained, through the primary basis of deformation. This is about the ways that naturalization are about the context of becoming and how becoming is about the ways that the world is about the nature, of man and God. God is about the becoming of naturalization too. The eyes of the world are about the conclusive evidence that God created the universe. This is through a way that we can understand things. This is why naturalization is there. This is an easy explanation of how the world could have been created. It is not fact. It is a continuous theory that we can learn. This is about how the nature and man have seen the

world. This is through the truth theory of the aspects of naturalization that are about the context and contextualization. This is about the real and imagined ways of nature and man. This is through becoming. The explanation of naturalization is about the process of being uncovered. There are coordinates in space and time and existence and being that cover the way of the outcome. This is how the world is created, and how the world is about perception, and the way that the world is about the outcome of the perception, that it is based on. The God of the world, cannot create the world, without a beginning too. Yet, the process of naturalization is a process. The real naturalization explains how the perception of the world, is about the creation also, as well as knowledge, as well as the circumstantial happening that created the world's becoming, that is about the ways that the naturalization is about its own naturalization, that is about the explanation of the ultimate perspective, or perception. This is about the context of the beginning that could have been Jesus Christ. In Christian theory of naturalization, there is a Jesus Christ. It explains that there is only one creator, who is the Father, and His Son, Jesus Christ. This is how a son would relate to his father. This is by also creating the world. The allegory of the doctor is this precisely. The doctor has a relation with his patient. The doctor heals the patient by also communicating and knowing, what and how, to do this, with philosophy of healing and making well. There are agents in the process of healing. These are pills and the substances and essences. These are knowledge and outcome, like the process of naturalization, is related to Jesus Christ. Jesus is the doctor, and the patient, is the person who reads the Word, and understands it, completely, with glory, and honor, and faith, and knowing,

what can come with His belief. The doctor prescribes the pills. Jesus prays and heals people.

THE ANALOGY OF THE DOCTOR WITH JESUS CHRIST-

The scientist and the doctor are about the complete, and total, lack and improvement, of faith. This is about Christ's body, and how it must be healed. This is by Jesus Christ. The doctor heals the patient, much like Jesus Christ heals the person in God. The doctor gets satisfaction. God gets a salvation plan of faith of belief for the faith that the person accumulates. The doctor uses knowing, and the Jesus Christ used the beginning. The beginning according to naturalization is about the process of healing. This is in the name of all. Without naturalization there would be a fall. The healing of Jesus Christ is in the reality of God. Like the doctor heals the patient, Jesus Christ, heals the man. One does it through science and the other, Jesus Christ, does this through naturalization. This is through the God of the universe, who heals through this type of thinking too, and in addition, also created this theory, without doing it wrongly and strangely like evolution. The doctor, known as Darwin, in philosophy most likely, won over the human race, to a theory that believed in monkeys that did not heal us. The Holy Spirit automatically heals us, so that the world is based upon the relationship, with doctor, and God, and patient and knowledge. The knowledge is still fallen, known as men. But, the redeemer of the world is always healing. He is the real God of Christian naturalization.

Comparison with Different Deformations-

The outcome of the philosophy of naturalization is the process of knowing. This is about and through deformation. This is about the ways that deformation is about and through the teaching of words. The process is the same as the formation of the universe, through the knowledge of what naturalization knows, of how it also evolved. This is about the opposite of evolution, this is about the creation and formation of thought, and thinking, and about how the thought and thinking, is about the ways that mankind, and man can relate. The outcome of naturalization is about the way that the purse, and the world, is for a woman, and the wallet, and the world, is for the man, with outcome of reality and the dream of man. This is to always understand himself, and always explain himself, and to think to exist, and know all of this without thinking, like a walk in the park. The outcome of the basis of the conclusive evidence is about the theory of the naturalization, and how we can explain ourselves, and know how to know and understand the theory. It is not a theory. It is a complete and total outcome, of what has already begun, and ended, and created, in this image.

The Outcome of Deformation-

The outcome, of how we understand deformation, is through the ways that the outcomes, have explained the processes of naturalization, and how we must begin, and end, and know about how it began and ended. This is through knowledge, of created, and unknown, created God. He is not explained as created through science, un-

less the real world of Jesus is involved. This is about the outcome of the explanation that makes sense in how the world is based out outcomes and conclusions anyway. This is for and about the conclusion that is always men. This is that the world had to be created by someone, according to man's rules and regulations that have always been around, and formed, by this creation. This explanation is a research and outcome of much anticipated thought and thinking, that is about and for, the world that we created, through God? This is how my theory works. Yet, I do absolutely know that it was created by Jesus Christ. The outcome of Jesus Christ was created by Him, which was created by God, which was created by a voice. This was considered "Ourselves" or Our Image, or something that is God. This is what God must have created through a false theory. Yet, all things scientific are false. This is a big question. This is about if 1. Science or 2. God is right. Science happens to be the one competing with this explanation. This is because it is the chosen art form of persuasion, and created outcomes, of God and how He exists. The God explanation is that He was a voice and He did not know how the voice was created, as well as Him. But, Jesus Christ came to save us, and He was also the real creator also. The Bible cannot lie or tell stories, but the world will see this theory. This is why I am scared. The science world is too vast and atheist that the conclusion I have drawn, might already be considered "insane". This is about the way speculation works. This is through strictly science. No other subject occupies time and space this way. This is through theories and explanations, with God. God is in the beginning, creating the world always. Yet, if man wants to understand the way the world works, then he must be challenged to face the world, and its counterpart, known as disbelief. If there were no Christians,

would there be only fake people? This means people created by themselves. This is the only argument of science. The basis of knowledge is about the recollection of where you learned it. This is either in school or in church. The language of deformation is about the complex and complete explanation of the universe. This is from beginning, to the end. This is about the way knowledge is about the essence and substance. This is about the all-encompassing knowledge that is in church, school, and the government, which makes up power and knowledge. These are the things we learn about, and think solely from, in essence and substantial beings. This is how the laws we follow, the belief, and the control of us originate and began and was started, and was created in a manner of this only. This is where all thought comes from, except for the evil of the world that makes us think differently. The houses we live in and the businesses and companies that we work for are existing because of these essences and substances, of control and authority and performance related substantial and essential philosophies. Without these things, we would be lost, and found, and lost, and found, and lost once again. This is why God created these institutions. This is because they were part of the evolution of our society. The naturalization is in schools. This is the explanation. This is the result. This is the outcome. There is no other essence. There is no other substance. This is of Godliness and His constructions and made essences, and substances, that are solely His, to be name. This is about the "naming" of the institutions and where they came from, and are inspired by. This is the "game" and how we "result" from the game. The "game" is described by me to mean the results of the world's stage, and education, and performance, which is the government, the schools, and the church. God moves only in the

church. Yet, in schools the church is always related. In government, the church is also where most go. Or, we end up in the result of the opposite for the rest of our lives. This is the lost and found. There is a movement in Oxford that is about the environment, and how we treat it. This is with dignity and strength. This is also an institution because of this. This means that the beliefs and core values of the school are about the trees and how we treat them. This is my example of other institutions. I think Oxford and its thought is the only other one subjectively for me, which I can fully comprehend, and know. These are the ways that institutions think, outside of church and government, with a conscience. There is so much power at Oxford that it is hard to imagine, a place, other than this. This is how God calls people. This is through the systems that have been established for us. When I leaned about trees, God called in me to go to Oxford University, in Oxford, England. This is where I will reside, because of this. This is about the way that the world is about the contextual analysis and opinion of God that is about relativity. This is a theory as well. I do like this theory. The coordinates of this theory are about the conflict and interest that are about the context that we are involved in. These are about the ways that the world conflict and are about the control and opinion that reflects our environment, with environmental interests, that does make sense, according to God. There must be a purpose to everything, and this is surely Jesus Christ. This is what naturalization is about that is about naturalization that is about naturalization that is about naturalization that is about naturalization, that is about the nature of man and its enemies. This is about the content of naturalization and its processes that are about naturalization that is about the reflexology and meaning of the issues that are about, and for

the essential needs of the world, and its contrast. This is about the need of the issues, and how the prevalence of the issues is complete, and naturalization is about this, through issues, that are these issues that are the issues that are the issues. These are issues of what complete and total issues, are about, that are about the issues that are about the issues, which are about the issue. This is of prevalence. This is what prevalence is about and for, that is about and for them from, which is about and for the issues that are about and for the importance of the issues. This is about the issues that are about the issues that are about the issues that are about the issues that are about the complete, and total ways of living, that are in the issues, that are in the contextualization of the issues that are the issue, that are from and for, the issues, that are from and about the issues, that are from and about the issues, that are from nature, man, environment, naturalization, God, and mankind. This is from the excellence of the issues, that are about the environment, that is about Oxford University, from which is about the places, where we visit, that are about the visiting, that are about the excellence, that are about the standard. This is always about naturalization. These are naturalization that are about, the naturalization, of man, and mankind, and issues of the complex, and complete way of living. These are the issues. These are about and for the environment. This is just my example. The way that the life of the God of the world is about the complex and complete ways of living, are about the issues from, and about the issues. This is about the ways the issues of the world, are about the stages of man, and how mankind, is about the stages. These are of the context outcome of naturalization. These are from the order, of thinking and thought, and complex and complete, orders of relevance that is of justice, that is of

the relevance, that is of thought, that is of thinking that is of thinking and thought. This is about the ways that relevance is about the progress of the order, of thinking, and thought, that is swayed every which way, toward the inclination of justice, known as the "laws of naturalization", that are really simple and good ways that we see mankind, of and for, with the issues that are about the context of complexities that are about the complex and complete ways of living, that are of naturalization, that are about God, and man, and about environmental complexities, that are about the context and content of the just ways that we see the illusion, and how the illusion comes to each person. This is about the ways that the illusion is about the ways that the illusion, that is about the dream, that is about the ways that are about the ways that the illusion, is, that is about the ways. These are of the ways of the laws that are of the world, that are of the ways, that the world, is about the way that the contents of context, is about the content oriented ways that are from the God of the universe. These are the laws of justice, in the universe, that are of the ways that the laws of the universe, are about God, and what we must follow, is about dictation and construction, of laws of thought, and the thinking involved in this and the ways we act. These are about the discipline and instruction, of God and what God is about, is about the complex and complete ways of understanding, that are about the God, of the world, and what we do to Him, with laws of the universe, and how we function is about the ways that God is about the context. These are the remote opinions that are of the God, that are of the content and firm construction, of the law of God. These are inside naturalization. This is about the laws that are of, naturalization, and what naturalization is about is about laws. The laws must be there to have construction.

The ways that the basis of thought is about the complete and complex ways that naturalization, is about the complete, and complex, ways of the laws, are about the construction, of man, and him. He is about the laws of him, that is about the nobody and somebody, approach, to naturalization, that is understood in contrast, to the law of a nation, and if you are allowed, by naturalization, to be a member. These are the laws of naturalization, that are about and of the context, that is able to recognize, the construct and reality that is of the laws of naturalization, and not just by the way we see the laws, of constructions, that are about and of, the content, of the ways that God, is about the ways, that God, is about the God, that is about, the ways, that are about the essence and substance. This is the law of self-communication.

The ways naturalization was taught, was through the dynamics, of justice, and the complete and total ways of naturalization. These are the naturalization of the mankind, that is about the man, that is about the, environment, that is about the context, that is about the concerning, that is about the will, that is about the way that follows man, and what he does, is about the nature, and man, and context, that is of and about, naturalization. These are of the naturalization, that is of and about, the context, that are of and about the relevance, that is of and about, the complete, and total harmony, between the parts of the theory, that is natural, in appeal and persuasion, that is of lies, and cheating, and stealing, and not about lies, cheating, and stealing, but about the context, that we are all inside. These people who lie, steal, and cheat own the world, most likely. These are about the content that is about the contention that is about the circumstantial happenings, that is about the complete and complex. These are the

ways of justice. There is no justice, when we are trapped, isolated, and confused. These are the containments of knowledge that is about knowing, that is about the context, that is of and about, the social adaptation, that is about the trapped and isolation disorder, that is always about our not understanding naturalization. These are about and of the naturalization that is of and about the people, who are innocent, also. These are controlled by the liars, and cheaters, and stealers, but are also controlled by the Jesus Christ. This is about the ways, that naturalization is about the context, that is about the content, that is always of and about the content, and contents that are complete and complex, scenarios, that must be adapted to, and come to, to be about and of the issues, that are about and for, the complex, and complete, unknown intelligences, that are of and about, the complete, and orderly, ways of mass confusion, that is prevalent when reading this book. This is all about the reflexology of the system and how it is about, the content, that is about the conditioning, that is about the relevance, that is about the God, of the world, that is about the contain, and complete, and construct, that is of denial, of the God, of the subject, and of the illusion.

The system, of justice, is always corrupt. This is due to the nature. The nature of justice, in the illusion, is the part of justice that is part of justice that is always a part of justice. These justices are about the content, of the illusion that is about the contention of justice and how its parts, are interrelated. These are interrelated, to the system, of corruption, and the laws of corruption. These are also, a part, of the naturalization, of the world. These are a part, of the naturalization. This is about the system that is about the systematic functioning, that is about the part, of the natu-

ralization, that is a part of the naturalization that is a part, of the naturalization, that is a part of the naturalization. These are a part of, the naturalization. These are part of the naturalization. These are part of the lying, cheating, and stealing that is a part, of the naturalization that is part of the reflexology that is in the Biblical interpretation of our world that is a part of school. These are the naturalization, causes of naturalization. These are the parts, of naturalization, and what they are about. These are about the cause and effect. These are how the cause and effect are about the complex and complete realities that are part of the system. The system is part of the dynamics, of the control, and conflict, that is always about our interest. These are about the interests that are lying, in deformation, that are about the control, and conflict, and resulting influence, on the apparatus of thinking. These are what cause inspiration. These illusions are the causes of the inspiration that are and are about, the conflict, and confliction of interest, that are about the ways that the living, is about the confliction that is about the conflict that is about corruption. This is about Satan. The confliction of Satan, with evil, and with corruption, is the downfall, of the systematic. This is from the result of Satan. These Satan, things, are the things that make up rules. These are about basic instructions. The rules of the, systematic, are in conflict, with one confliction, and another self interest, that is about the way that rules are about the game, that are about the confliction. These are about the colossal, ways. These are of thinking, and acting, and socializing, and are about the ways, that we rule our world, through laws. This is also what Satan can't get to. This is because he is an angel. But, with deformation, there are leaders, and there are followers. The ones who create deformation and the ones who agree with it entirely. This is

because of existential, ways of justice that are from corruption. This is a contest of the survival of the fittest. The fact is that the essences of the first and last proliferation are about religion though. The fact that the proliferation of the justice, system, is about the collaboration that is used within religion. The fact that the religion of the world is about the context and content and conceptualization that we are preprogrammed to do, is about the world, which is about religion, in a small scale. Comparatively, there is a hierarchy of religion in the ways that we view the world. This is about the contrast and comparison of the world that is about and through the context that is of naturalization, which is of evolution. But, therefore, in evolution, we have no value. That was the point, to educate all. But, in a way we didn't like. This could be for survival of the fittest, which is false. The false evolution theory is a creation of many things. These are the complete and total package of 1. Apes, 2. Competition and 3. The creation of the world. These are about the true nature of God. Yet, this is false. This is about how false things, rule over the world, and consist of details. These are small lies. The theory teaches us that we are all equal, and the equality of human beings, is created. Yet, this is by a big bang. This "big bang" is about the exploration. This is of where we go. This is to school. This is also to the forest. And, this is also to lies. This happens over and over again. The reason for this, is moot, therefore the calling to exist with the big bang, is not existence, but trust, in a higher power, that is not there. The basis for this argument is that the argument itself is already not true. This happens over and over again. This is until you believe it, because of the way it was written. This was not written for Christ, but for Satan. There is a Satan, it was written for, and this is you. This is if you believe in it.

This is just an excuse for evil. This is from the very bowels of Satan. This is the holy theory that is actually pure evil. This is what pure evil does, that is about pure evil. It makes you believe in it, until it is true. This is about the ways naturalization is true. This is in the ways that it is natural. The belief system, of naturalization, is always true. This is because it is relevant. There is no God of evolution, but there is a natural God of naturalism. This is the natural.

The ways that God, is about, the earth, is about the ways that Christ, is about Him. The body of Christ, is about the surviving, that is about Christ that is about naturalism. This is about the Holy Ghost that is about surviving, but not in the context of evolution. One is a lie, and one is true. Naturalization is true, and evolution is a lie. There is certain knowledge, faculties that know about Christ. He is in naturalization. Yet, He is not in naturalization that is from evolution type of thinking. He is the sole person involved, in surviving, that is true. This is about how it is true, and it is about what is true, and what is false. There is a false, notion, of independence, that is within evolution. This is that there is evil. There is no such thing as evil. Yet, with evolution there is? No, it is just a lie, or a false creation, that is not "evil" but "right" somehow, with the Holy Ghost. There is no evil, in naturalization, unless you choose to make it this, exact way, with your beliefs, yet Christ is there for anyone who believes in Him, as God. He is not evil. These ways that evolution, works its ways, through the evils of the society, are inside fact and fiction, that is irrelevant and implausible, and incorrect, in aims and ways, that we are functioning with, and from about. This is about the simple ways that are about the pure and simple. This is about the ways that disbelief and belief

plan a role into the world stage of naturalization, that is and are about the dynamics, of change and reoccurring ways, of vision. This is about a complex way that nature and naturalization are about the extremist and occurring ways that naturalization is about the ways. These are the ways that the world views extremist and opportunistic ways of violence, and sex, and drugs. These are, for example, mind-altering realities that complete. They do not gain though, but repeat. These are why the notions of the simplistic are about the complex and complete ways of math, and science. These are about the roles of the true and untrue. Naturalization is my theory, and what it says I do teach. The unresponsive, theories, of naturalization are about the reflexology, and the dynamics, of which thing, or nothing, accepts, this demeanor, and challenge, and outcome. These are the details of a master of the universe, which is control. These are of the real Jesus Christ. These are the details of what the contentions of the complex issues are completely, and concerning what are about and make complete, with essences. These are essences, that are the everlasting, and outlasting, of the even keel ways, that are about fiction, and fictional, ways that are about the setup and complete. These are the dynamics of what naturalization is really about.

The Complete and Total Ways of Dynamics-

What I am telling you about is about the instruction. This is about the false ways that evolution excels and nature demands. This is through the real Jesus Christ. These are issues about and for the face of the universe. This is what it demands and controls and explains. These are holy.

These are right. These are complex. The real Jesus is completely about the real ways of communication, through the aspects, of existence. These are the details of the existence. These are relative. These are true. These are of what Jesus Christ, is about, and prevalent to, through explanation. He is the teacher. He is the class and the instructor. He is the God that is about the design of the universe. All things are special and relative because of Him, including justice. He is the Maker, which means that He is solely responsible for us. But, all people are made up of Jesus, existentially, which means He is made up in God. He is the Son of God, who is the Creator, of the universe. The ways that are about, the relation of God. These are special, because we are all special. These are relative because of Him. These are explanation worth because of God. These are worthy of any man judging you. These are worthy of the special relation with the essence, and substance. These are of man, and His God. We are the people, and He is the strong person who knows us. These are about the ways that God created the universe, and how He created this, is in the images of God. He was the man that He created. This is with His. He is the creator. These are about the ways that man, and nature, and naturalization, is of the makeup of the close relation with the Maker. Darwin considered himself, the maker, though. This was false. The theoretical faith of mine is about the consistency of God, and the reality. This is of the intellect. These both are theoretical. This is according to me. Man, and God, and man, and God, and man, and God in control these relative aspects, of God. These are all connected by a relationship. This is with the Devine. The aspect, of man and God, are critical. These are issues that must be related prevalently. These are issues that are controlled by God. These are because of Him. These are not my theo-

ries, but my intellect. This is part of God. The way God is about naturalization is about the land that you are naturalized in, comparatively speaking about God, and how you then have a home. This is where you can go to. This is about the way the land, where you go to, is about the in toward. This is of God that we go towards, and then go to. This is unto where the God of the world, goes unto, and makes a disciple reach one other individual, with essence. This is how God works. This is unto naturalization. This is about the way the nature, and man, and God are about the way God works. This is about the way God works, through God, that is about man that is about man, that is about God and man. Absolutely, God is about God and man. This is absolutely. Yet, some other variables go into God and the theory of naturalization. This is about naturalization, and how the theory, of nature and man, also work out, through naturalization. This is about the comparison of the world, to God. These kinds of people are about the ways that God and man, work. These are work. These are performance, and the way God is about performance. This is about God, and the way God, and man, are about each other. This is about the translucent nature of God, and man, and how these variables work out. This is through the faith of man. This is about how God and man work. These are for God and man. These are about the way of God and man and about how they work, are about God and man, and how they work. The way that the agents of God work, are through God and man. One must know about God. God must be alive and real. Man is about his relationship with God. This was from the beginning. V This is about God and man and how the fellowship of man and God are about the translucent disguise of man, and how he must have sin nature. But, this can be created

out of him, when God creates it out of him. This is all about Christian relativism.

BASIC NATURALIZATION WITH A CHRISTIAN BASIS-

This is only scientific. This is the outcome of naturalization and how it is decided, on to be. This is about the ways that merely science achieves, and completes, the naturalization. This is about God and how He does create naturalization, and the way that we think. He is about science. God must have created science. This is the formation God, who created the universe. He is about the essence, and substance, of becoming with man, that is only God's. This is truly because of science.

The outcome of science, is always antireligious? This means that God has nothing to do with science, in the way I just explained it. The scientific method is man proving himself to God. This is, yet, without words. This is just a dry bone. This bone has not been picked, but has been placed in his heart, where he cannot see anything but his shadow. This is where his bone has left him. He is in the shadow of God, and the bone is now gone, in another person.

The dog is about the theory of naturalization. This is in the same way. This is because I am not saying I am God, but a direct outcome of the process of thinking, and where it gets me. Naturalization is what I had to write, to explain my explanation of God, and how I cannot fathom a creator, without seeing what He created, and "became" with. This is a simple theory. This is a math equation, followed by a

scientific explanation, followed by a physics following. This is what naturalization God relates.

The theory of Naturalization-

(Equation)-

This is the part where the theory goes.

The evolutionary theory of naturalization and how it makes sense. This is actually anti-evolutionary, in essence and substance, of God and opinion, yet it has a piece of information on the survival of the fittest aspect, that is of natural man and himself. This is actually a piece of work, that is about excelling and surviving also, which is about the ways that the dynamics of justice and the promotion of it, is good.

These are the dynamics of the justice, system, that is about and according to survival of the fittest, and about naturalization. There is a feel, for justice, and the dynamics of surviving, that are and is about, the coordinates of the third section of the theory, in math and science. There is a strong tie to the world, that is about the naturalization of man, and how the naturalization, of man, is about and for the justice, of man and himself. Yet, even though justice is about God and man, the survival of the fittest, is about the justice of man, and himself. These are the own opinions, of naturalization, and the way it treats its subject matter, in case and form. This is about the ways that the nature of man and how he is about and for, the subject, of man, and himself, is always about nature and man and his opinion, of his own justice, which is also controlled by God. The ultimate creator of the universe, has

to be about God, or we would not have been created, by God. This is about the ways, in which, control, and justice, are about the dynamics of the whole theory, and about how the dynamics of control, are about absolute power, and how we must have an explanation, of the power, that is about us, within the context of understanding. These are the issues of naturalization, and its standings, that are about the absolute power, and the fictitious "ultimate truth", that does not concern this aspect of naturalization. This is how justice, could be only created by God, and about how the naturalization, is about man and himself, and about how God, is about the creator God, and how the man cannot create him.

So, in conclusion, therefore, that this is what is always and is for and about the natural God, of the naturalization theory, which is about justice. This is about and yet about how the conclusion, is that man cannot comprehend this aspect, of the real God. These aspects of God, is about the naturalization theory, and how the man of God, cannot explain this creation origins through the origins of academic and social and religious justice, due to the creations of the world. This is because it is God's. The God planning, and outcome, of social justice, is relative, and complete, and is his origins, with naturalization justice, which is the God of the world's, and His religious Plan's, that are and is about, the religion, and is about social justice. The apprehending, of the apprentice, and his tools, are about the conclusion, and justice, and reality, of the justice system, and its appeals to the justice system. The God of the universe, is about the social, and academic justices, that are these, that are of and about, the conclusions and philosophies of the world. The facts about justice are the whole and nothing but the truth appeal and dynamics of

controlling the world, through laws and legislature aspects of the truth. But, where does the truth come from?

The real absolute truth is about the aspect of the humanity that is not in man. This is in the God of humanity. The aspects of human resources are of and about the knowledge, of man. This is about the aspect of mankind and man that are hidden in the truth that is of God. These are the aspects, of truth, that are and is of and about the way that the hidden truth, is of the God of the world, and where He chooses to go. These are the interrelated aspects, of control. These are of the written words that are of God's, but of the Biblical interpretation.

This is of God's. This is about the aspect, of truths, that are and of about, the essences, of truths, These are the aspects, of the hidden truth. These are Bibles, that have aspects, of the truth, with an hidden agenda of truth, that are of the Biblical. These are of and about the truths, of the Bible and its relativity to the truth of Darwin, and such. These are the Biblical relations, of God and man, to the whole truth and the absolute truth. These are of society, and are of and about the real truths, that are of the Biblical versions, of the truth, and where we go to find out about these, which are not on earth, but are in the control, of Godliness, that are in and about our individuals, that are and is about the controlling of the earth. These are the aspects of truth that are, of and are about the essences and condemnations, of the essences and substances, that are of and are about the philosophical, method and odd way of performing. These are the laws of the real justice system. This is about the aspects of justice and the realms that are the performance of justice in the eyes of the people, who have made the laws and are about the ways that the laws behave and perform. These are of the

eyes that are about, and of the, ways that the control and manipulation of the system, occur. These are the simplistic ways that are about the complete and complex ways of justice, and the creation of the realm that he is in. This is the fallen man, and its counterpart that are of and about the complex and complete ways, that hate is bred, and natural things occur, that are of an illegal nature. This is about the ways of man. These are of and about the nature, of man, and of mankind, that is about naturalization and his kind of thinking, for here and abroad, details of the man and mankind, in the opinion of the outcome of God, through relevance, and the intelligent design. The factual existential mankind is about the knowledge that is found in the research that man proposes, to himself, through his own senses. These are about the senses that are about the senses, These are the conflicting interests that are about the dichotomy of good against evil, in a manner of speaking, that is truly true. These are about the things, that make up the world and are about the special dichotomy of good and evil, and how they are seen. These are what is truly seen through good and evil. These are all the senses see through, absolutely and subjectively, that make up naturalization. These are naturalizations that are about the senses. These are about the phenomenology that is about the future that is about the past. This is about how we deal with good and evil. This is the actuality of the theory of deformation, that lies, cheats, and steals, but doesn't. This means that the evil part is gone, unless you do not agree with it, wholly. The evil, and good, are about the senses and conflict, of the senses, that are about the conflict, and the interest. These is the outcome and goal of naturalization. These are about the naturalization of the world. These are ho w good and evil see this phenomenology, to a exact science. These are about the things, that

are about God, that are about man. These are the "knowledge of good and evil", which is solely thought. This is about thought that is about exact measures, that are about the law and how we must have it. This is because of the way I am arguing, because of it, because of the makeup, and because of the world, that exists, to every individual's thought, and directed proposal of thinking, that eventually catches up to him. This is his perspective of God.

This is about the way the world is about the conflict of interest that is about the senses, that is about research, that is about knowledge. This is through the knowledge of God, that we all have, through the fall. These are about the senses. These are about good and evil, and knowledge. This is about how the senses, view the world, through the fall of knowledge, in an absolute or subjective calling. The answer to the world view of policies and procedures, is about how the world, is about the lesson, we learn, and how we learn this, through the nature of man, and his environment, which conflicts with everything, unless he is good and evil. These are the nature of evil, in which the nature of evil, is about the conflict of interest, that is about the good and evil, that is about God's that is about the senses, and how they now have knowledge, and this is in after the fall. These are about how the fall of man, is about his naturalization, that was the way the world saw, after the fall of man, because of knowledge, that is comprised this way, and contains itself, this way. Essence, is about the generated, theory, of knowledge that is about, the conflict, of interest, that is about the sense, that is about mankind, that is about man, that is about the knowledge, that is about the senses, that is about the world, that is about the senses, that is about the God.

This is about God, and how the world that is about the way the directed outcome, is about the ways that the view of God, is always afterwards through knowledge. This is all about the confliction that is in the best interest of man. This is the curse. This also contributes to the knowledge of the world, which is the laws after the fall. This is about the ways that man's perspective is about the ways that man is about the way that man is about the way that man sees. This is through the naturalization of God, and how man and God, can be about himself. This is through the eyes of the religion of Christianity, and how its theology plays an important role in individualism and about how the individualism, rules out evil and good, and makes itself naturally prone to the basics. These are the basics of knowledge. This is because Jesus Christ is God's Son. He is the one who came to save us. This is from falling again. These are about the conflict, and interested person that is about saving, that is about complete and comprehending the totality of Jesus Christ. He is God also. Therefore, in Christian naturalization, there is a maker beside the Creator of the world, that is about sin. This is about how sin is also part of naturalization. This is intense, because the ways that naturalization is about conflict, eventually proposes a offer. This is a free gift. Yet, Christian naturalization is always about Jesus and His plan. The creation theory of naturalization sees everything in its own image, which is God's until Jesus came. These are the facts of naturalization and how the facts, are about the progress, that are about the reality that is about the realization, that is about the becoming. This is about the sin, of the world and about how the sin of the world, is about Jesus. He is the Master of this. This is His kind of knowledge. This is Him as a judge. The world is not prepared for Jesus Christ still. This is why He is prepared for anything. This even-

tually becomes His, and He creates and destructs, the world that is about the naturalization, that is about the God of the universe, and how we are related, to Him, through knowledge, that is of His, that is natural, that is all from God. These are the gifts, of God, that are related through the essence, and substance, of the good and evil, and about how the naturalization of the world is about the essences and substances, that are truly its own. These are what naturalization is created by, and in its entirety completed with, and for. This is for the real God. This is about the way that nature and man, has harmonious relations with each other. This is from Jesus Christ. The maker of Jesus Christ, is about the naturalization, of mankind, and about how the nature of man and God, are about the sinning of the world with salvation. This is through the evolutionary confliction as well and is about the conflict of the theory itself with religion. These are about the confliction of interest that are also about naturalization, but with a Christian context, of believing also. The belief of naturalization is about the senses that are about the confliction that is about the conflict of interest that is about the senses that are about the confliction that is about the senses that is about knowledge and how the world knows good and evil only. All thinking eventually goes into the world, with thinking that is about the senses that is about the sense, that is about the evolutionary context, that is about losing, that is about the confliction that is about the sense that is about the creation of the design of man, and how man and his own conflict ion are about how naturalization was created. This is from the fall of man that is explained through the theory of the math portion of naturalization. This is about the senses and how the communication through the sense, is about the total, and complete, ways that mankind, is about the conflict and interest of

man himself, which is only explained by sin. The extra ways that sinning is about naturalization is through Jesus Himself. He is the redeemer of the world view of sin, and about how the world view of sin, is completed and judged, for the Garden of Eden. There is a wonderful excellence to the world design of naturalization, that is about the ways that naturalization is about the process of naturalization that is about the senses, that is about the control and authority of the church. This is the "body of Christ". The naturalization of the world is through the nature and man of man, which is about sin nature, and Jesus. This is about the ways that Jesus Christ, can explore and become with the explain and analysis of the completion of the project, that is of and about the controlled outcome, that is about himself, that is about man, that is about mankind. Jesus Christ, is like naturalization in the ways that we see Him. This is His naturalization also that I cannot explain. This is the way He has knowledge. These are the ways that naturalization and man, are about the accompaniment, which is about the conflict and construction cause of intelligence, and how Jesus Christ is God, so He has something different. This is also naturalization. The real process of naturalization, is through the explanation of the goodness, that is with man and his complications, that are about the sense, of becoming, and about the ways that becoming, are about this, are about the confliction, and about how they are solved. This is all based on knowledge from a simple equation. This is about the context and confliction of the fact. This is all opinion based or fact based, but all it is, is a perspective. This is from the illusion or reality. The creation of the world, through naturalization, is through the eyes of the world, and its basis, that is complete. This means that the world, can see through the eyes of itself, and know what it sees. This is about the

way that the world, is about the way that the world is about the way the world is about the way the world is about naturalization, and how the world is about the naturalization, of the world that is about the good and how the good is about the nature of God, and its environment. These are how the worlds of naturalization work. These are through the details of the justice and environmental thinking. This is about the ways that the meek work, and about the ways that the meek inherit the earth. This is about thinking that is about the essence of man and what he provides to the world stage, and heritage of man, and his thinking. The thinking inspired and contributed to this world of thinking, is about deformation and naturalization, with impulses and relativity to the strong and gentle Jesus. He is about all knowing phenomenon that concerns itself with the knowledge of God. He is the Omniscient, Omnipresent, and Omnipotent creator of earth. He is also the creator of Heaven known as Christ. Yet, the world was created by a voice, and the Word was always in the beginning. This is my basis of explanation to the theory of naturalization. These are the theories of mankind, that are always true, in the special way, that we see and view the world, through naturalization. This is about the creation of the world, that is in the mind of the creator. He is the one who created naturalization. This is about the ways in which naturalization is about creation. He is the one who is about the ways that naturalization is about the creation. He is about the way naturalization is for the creation. This is in terms of God, the maker of heaven and earth. This is about the way that naturalization is about the prosperous and genteel notion of God, and the approach from people. This is through real knowledge of good and evil from God. These are promotional issues of the environment, and the issues of the context of content

and the research that accumulated into the knowing of the world issues of relevance and the issues of God and man that are about the context and about the content that is about the essence and substance of God, that is about and from the essence and substance of the man, in the world of the relativity of the God. He is the one who created us this way, until we fell, into the abyss, of God's very own knowledge. The evolutionary theory is sound, but doesn't make sense to me. This is about God's plan and how this makes sense to me. I do not know what a lunatic Darwin thinks. This is about space age thought. These are about the missiles and firearms that are about the evolutionary theory, and how this makes no sense. It is like seeing a missile and thinking about another missile totally different in the same way, because of the way he tricked our thought, supposedly. This is another issue of thought, that is about the basis of thought and thinking that is about our nature and man and how he can become man, within the prodigy of man. No monkey could do this. And, this is about naturalization and how this issue is made sense to everyone. This is about the monkey and how it does not have reflexes. This is through God. This is about God, and how Darwin also had a God, known as his book theory that enveloped him too. This is about issues with faith, and how God created the world. Darwin slaps me in the face. Yet, I must forgive him. This is the complex way. This is about Darwin and how he has an ego. This is about lies, and falsehood. This is what Darwin was king of. I expect to be 100% better than this kind of thing. He is a copout of the modern day world. This is with meaning and purpose, that Darwin could not handle. This is about Christ, and how he is the false God. This is why. People like Darwin make him out to be this. Yet, he never gives up. He is about the language that is about man, and the

mankind, ways of living life, with man and purpose. He is about opportunity, and anything he wants to be. This is even a monkey. This is if Jesus wanted this, This is if man wanted this in Christ. This is why Darwin is Christ, in a false way. He was too smart for evolution. This means that his theory was false. This is evidence of the Christian way. This was as smart as God. This is because Christians are God. They are part of Him, where non-Christians are not. They are part of the universe. They are separate entities that are, are how they say they are to be. This is not a lie. Yet, Darwin lied throughout his theory. This is because of apes. This is why apes aren't Christ. And, the biggest part of church is truth. This is about a play in football. This is the touch with the down, set, hut to the man and throw play. This is from the running-back to the quarterback, after the throw has been executed, to the running-back safety. This is about the winning of the game. This is not a Darwin unless there is Satan. If there was no Satan, there would be no Darwin, and if there were no Satan, there would be no evil. This is why football makes sense to me also. This is because it is dominated by glory, honor, and pride, that leads to successes, instead of failures. These "failures" are against the system. They are about the play. They are about the execution of the winning. Satan does end up losing, This is why maybe we are in competition, that is ours. This means that we are not deceived, but tricked by the real world, that Satan does controlling to, and ends up losing with, to Jesus Christ. As easy as you can do a football play, you can become a Christian. The way that Satan wins, is through losing. This is through the end. The end, is about the end. This is about the judgment. There is a seat in heaven that is about a judgment. And, it is, ironically enough, the real friend of yours, Jesus Christ. The end of Satan, is about

the ending of the world, with the return of the Garden of Eden, which is why Christ "wins", over all people, even the corruption in the world known as the "police". This is because justice is thwarted, due to the "world", (or Satan's endless plan). If Satan ruled the world, it would still be God's in the end. This is solely because of Jesus. The end of the world is not coming "soon", but is inevitable. This is like a game winning football play, that thwarts the justice system, into winning, over the attitude of the sports members, and ending up in a winning, establishment. This is about school. This is about all things that are about the winning, of the losing, and of the satisfaction that is always there. The winning is about the complete and complex way that naturalization is about a God theory and how the naturalization of the world is about the God of the world, and how there is a God, of the world, that is all about the complexities, of justice and how the justice is about the way that the justice is about God. He is about the all-performing aspects of Jesus Christ, and man. These are about the ways that God is about the world. These are about knowledge and disorder and order, and the complex and complete ways that God exists. This is about the way that the world is about the completion of the world, that is about the God of the world with the maker and how He makes the entire world, of essence and contemplation, that is about the goal, and what the goal, with Christ, is. His is a perfect world. This is about the religion and how it is a obvious aspect to my life and living times of God. He is the perfect being. God is about the ways that God is about the ways God is about the ways, in which that the real God is about the ways He survives. This is through church, and the government, and the school system. This is also about the business or "real" world. This is about the ways that the world is about the

ways that the world is about the way the world is about the way the world is about the way that the real world is about the ways that revenge is not right, and principles of naturalization is about the truth, and how it hurts the God of the world, until the beginning was the worst sin ever, made by a snake, a person, and a error. This is the trust God had until Eve ate the apple. This is about the truth that is about the way the truth is about the way the reality truth is about the spiritual truth, that is about the whole Bible. This is about the way the truth is about the way the truth is about the way the world is about the God of the world, and how we know Him, through truth, that is hidden and transparent, that is about the philosophical truth, that is about Young Life, K-Life, and Campus Crusades, and the churches that we go to. These are merely, a science, that has yet to be discovered. This is by Satan. He is the angel that cannot go to church, and cannot go to school, and cannot work. This is because he is an angel, that has only been, in life. This is because, he is the angel of death, that is about the way that the angel, is about the revenge that is about our universe and how he is alienated like a alien, to the religion of God. God does not see Satan, or he would go astray. This is merely with perception. This is about how intelligence and perception, is not a lie. But, Satan is a liar. He is the way the world sees evil. He thinks he is a king, and his opinion of himself, is the best ever. This is opinionated successful go insane, against people, in a split second, of time. This is how he fell from heaven according to Bibles. These are not the judges of him, yet some learned people see into the future, and the past, through issues, in the real Bible. The real Satan is minded on challenges and successes, according to me. This is my opinion, on the best. Yet, he fell from heaven, so he is the worst. The way that the world, is

about Satan is about the ways that He, is about the ways that Satan is about the God of the world, and is about the ways, that he, the Prince of Darkness, who is also a weird creation from the world, even through Bibles, controls the rules of the world, and principalities. If we did not believe in Satan any more, then we would be mindless, and worthless, and scared, because he would take over the world. Yet, we are like this anyway, because of him, anyway, meaning it is already that way, due to evil, and the world, and issues of evil, and how the real is only evil, and the fake, is only great. This is a result of Satan, maybe, but man also has knowledge that is from knowledge from the Garden of Eden. These are the ways that the man, is about the God, that is about the man, that is about God, that is about Jesus Christ. He is the Prince of Darkness, that is only about the escape, of knowledge, and what is good for this. This is about the way the world, is about the way the world, is about the concerning knowledge, which is about revelations, and the ways that, man, is about the fall. This is about the way that colossal knowledge is about the reason, that we have existence, which is about the way that the knowledge is about the God of knowledge. This is how the world was created? Try, naturalization explains why we fell, through an equation. This is also about knowledge, when we fell, if we can relate it. This is relative. In this context, relative, means, that the world, is content and happy, with the real and not the fake. The fake, is not natural. Satan is the fake. God is not natural either, but He is real. The fake verses the real is Satan verses the real God, which he tries to be. This is an obvious conclusion that is about the dynamics of knowledge of good and evil, and how they work, through God, and nature, and man, and mankind, and naturalization, and environment. This is about God, and how He has a language

too. This is about the ways that God is about the ways that God, is about the truth and is about the ways the truth, is about the God, that is about the goodness, and how we survive, with survival of the fittest. This is about the content that is about the contextualization that is about God. He is the God. He is the real Jesus Christ. And, He is the body of the church. No one, could go unto God, until He is forgiven for his sins. And, this is by the real Jesus Christ. Naturally, God created the world, with a theory, that I have, because of the way human knowledge does work. This is about and through the dialect and knowledge, of man, that mankind, and nature, would see this way. The nature of God, is about the human, being, that is about the challenge. This is about the way, that the humanity, of God, does do it all, in history of time and the world. He is a prophet, with many people, in the Spirit of God. I am not saying that I am a lunatic, but I am saying, that I am true, and truthful, about the Scripture, and the Bible. This is holy, and has always been around. These are the conclusions of naturalization, that I have always had. This is my perception, on man, and how he sees through God, and His perspective. He created the world, and has a plan, and can do the things, that people go for, in essence and substance, which is about God, and Jesus Christ. This is eventually, and for the world, that is about the way the world, is about and for the God, of the universe. This is about the conclusion, that God is Jesus Christ. This is about the way the world, works that is about involvement and how we have conclusions, that are about and are for a, conclusion, that is all Jesus Christ's. These are principles and parables, of God, and Jesus Christ. This is how we all learn. This is how we all think. This is when we are Jesus Christ. This is about the ways that following is about the way, that following is about the

way the world, is about the content, and context, of the being. This is about the Holy Spirit, and how it knows and discerns truth, through the Bible also, in manner form. This is manner form, meaning that the world is about the mannerism that we live in, and beside. This is our lifestyle. The reason why we have a lifestyle is because of the element, of desire, that is in the fact, that is in the opinion, that is in the religion. If Satan was fact, he would lose, and if Jesus Christ, were opinion, then He would win. This is internally how they are. Jesus judges, through His opinion, and Satan, decides through factual appearing things, like magic. The ways that the world was created, is through opinion, that lasts forever. There is also a free choice, that we are given by Jesus Christ, once in our lives. This is to be a good or saved person, or a bad and evil person. This is about our opinions. These are the ways, that man and nature, look past, the presence, and look into the world of the being. This is a being that has always been around, and has concerned itself, with the challenge or loss, of opinion and gain of Jesus Christ, that has been made sense, of the utilitarianism, of Jesus Christ, that is about the way the world, is about the way the entire world, that is about the way the world is about the challenge and execution, of sin. This is all through the entire world, that is about and has been, about and around, the challenge, of fiction and fact, that exemplifies itself through the challenge, that is about judgment and "turning the other" cheek. This is about the way the world is about the entire world, that is about the nature, of God, that is about the opal, of fictitious origins, and how we do not know this because we have not witnessed the birth of Jesus Christ, but through the faith of Christ's body. I think this way because I am a good person, who is a theorist in Christianity, especially with my thought, my whole life.

This is about the way that God is about the ways that God is entirely concerned with the world, and what the world is about is about the option of God, and the reality of Jesus Christ. He is the opinion, with the fact, that God created the world, spiritually. He is God, and we did not know this, until He came. Yet, there is a God also in Heaven, that is also God, in which is learned in church. The obvious is that 1. Christ is God, 2. God is omniscient, omnipresent, and omnipotent, and 3. The body of Christ, is about all of these aspects, that I have mentioned which is pure fact, and 4. The opinion of God is what shapes us with these collectivisms, in which shape Him, as well as our own selves, and are faith-driven, and chosen, through the church, and 5. Christ is the body of Christ, and His Plan of Salvation, is always prevalent and strong with faith, that is true, and what we think is His, if we are in Christ,, and 6. The God of the world is about the science and concerning of and about, the truth and what sets us free, and 7. What the form of God, is about is about God and the relevance of Christ to God. These are all about the ways that man is about God is about the way God is about the diction and word choice, of His also. This is about the way that God is about the concerning that is about the ways that man is about the way that man is about God is about the way the Lord is about the way the world is about God and what God is about is about the way that God is about the crying, broken, and beaten people, who are also about the complete and total aspect of God, and what God is about is about God and His presence. This is about the God of the world that is about the God, that is good, and is about the way church is about the way the world is about the way the world is about the God of the world that is about the collective and aspect directed outcome of the survival of the fittest, and how God, is about this and about the

church body also through the church. This is obvious that we have to believe in God to be saved. This is about the way that God is about the complete and total, creation of God, that is about Jesus Christ, and about God that is about the creation of God, and the real Jesus Christ. This is about the complete, and total, ways of understanding, that are about the 1. rules, and 2. How we follow them. These are about the collectivism that is about the case, that is case sensitive, that is about Christ, and about how the world does rely on Him, that is through the direction of good, that is about the direction, of God. This is about the secrets that are about God and how the world is about the real God, that is about the makeshift philosophies that are not about the world, but about the science that is about the strange and unusual things that are about the science, of God. He is about religion solely? Maybe, but most likely not true at all. The order of naturalization is about the way the naturalization of the world, is about the complex and complete ways of thinking. The reason why we complete and complex the naturalization is by nature that is about the ways that naturalization and thinking rule the world. This is about the way that the naturalization is about the process of health and concern for the wellbeing of man. This is about naturalization and the issue that it faces. This is about the way the complex and complete ways of naturalization are about the essence and substance, that it fits into, and completes, with major detail, of the naturalization that caused it to happen in the first place, which is complexities, that just look normal. This is about the ways that naturalization and itself are contained in the relevancy that it trusts, within issues of health and concern. These are about the ways that naturalization are about the concern, and concentration, of the details that are of the approach of the naturalization, that

is about the complex and complete ways that naturalization exists in, and is about. These are naturalization that is about the issues that are about the ways that nature and man exist and how they exist, is through the existential ways that philosophy and matters of psychology, are always about the existential matters of mankind. These are about naturalization. These are about the related issues that are about existence and about the ways that existence are about the ways that existential matters are related to existence. These are existence and what existence is about is about the control of the environment and the animalistic instincts of nature, and how it is sin related in man. Therefore, naturalization is true. The fact about naturalization is about the fact, that the fact, is the fact. This is about the way the fact, is about the fact, that is about the way the fact is about the fact, that is about the factual existence, that is about the fact, that is about the fact, that is about the fact, that is about the fact, that is about the fact, that is about the fact, that is about the fact. This is about the way the fact, is about the fact, that is about the fact, that is about the fact, that is about the fact, that is about the fact, that is about the fact, that is about the fact, that is about the fact, that is about the fact, that is about the fact, that is about the fact, that is about the fact, that is about the fact, that is about the fact, that is about the fact, that is about the fact, that is about the fact, that is about the way the fact, is about the method of fact, that is about alternate theories of naturalization. This is about the fact, that is about the fact, that is about the fact, that is about the fact, that is about the fact, that is about the fact, that is about the factual existences, of different realms of naturalization, that is about fact only. This is about the facts, that is about the facts, that is about the facts, that is about the facts, that is about the facts, that is about the

facts, that is about the facts, that is about the facts, that is about the facts, that is about the facts, that is about the facts, that is about the facts, that is about the facts, that is about the facts, that is about the facts, that is about the facts, that is about the facts, that is about the facts, that is about the facts, that is about the facts, that is about the facts, that is about the facts, that is about the facts, that is about the factual existences, of different realms and mediums, that are about the existences, and how they are interrelated, that are about the factual existences, that are about the factual existences, that are about the factual existences, that are about the factual existences, that are about factual existences, that are about, the factual existences, that are about the factual existences, that are about the facts that make up the factual existences, that are about the factual existences, that are about the one and only, factual existence, that is about lies, cheating, and stealing, that is about the way, and the way the truth can alter. This is about how the way that the world, is about structuralism is about the complexities, about the ways we can see our world, and how they are different than the ways that we can alter, and explanation wise capture and imagine, a world beyond ours. There really are alerter realities and fictitious realms, other than ours. These are naturalization. These are the man, the mankind, the nature, the environment, the naturalization, the God, and the process of thinking, that is about and of man, is about and of the nature of the man, and the naturalization, that is about the process of man, that is in thought. These are the thoughts that are about naturalization, that is about thought realities, that are about different realms of thought, that are about and contain, naturalization thought, that is about the essences and substances, of thought and thinking, that is about the ways that thought, and the ways of the way that are about

thinking, and thought realities, that are about the supremacy of thought, and how this is how the realms of thought are about naturalization, and about how it formed them. This is about the ways that thought are about nature, and man, and about how the nature, of man, and nature, and the man's nature, and the nature of man, and the way of mankind, is about the mastery of the realities, and about how they cannot alter to be another reality, beyond itself, in thought alone, and essence and substance, that is always right. This is not just another reality, but a thought realm beyond other realms of existences, according to liars. According to cheaters, the reality is theirs. According, to thieves, there is another reality beyond ours, that is really our own, that we cannot master or perfect, unless it is solely ours alone, but also, and in addition, to this, after it is solely ours, in alter imperfection. These are about the ways that the knowledge, contained in the equation, of essence, and substance, is absolute and subjective, and how the thinking and knowledge, is and are about the knowing, and showing, of listening, and controlling, is about the ways that we have knowledge, and how it is used, and practiced, which is in whole and in part, through the real universe. This is about the knowledge, that is of the other order, that is of another calling, that is of another context, that is solely the creator's. These are the creators that are the masterminds of the universe, and where they go, all others follow. These are the masters of the real world. These are always through thought and thinking. Naturalization, is about the pioneers of the new world. These are about the condensations, that are about the reflexology, that is about the contextualization, that is about the contextualization, that is about the authority, and power that is about the context, and contention of the well respected people. These are also ones who have

mastered naturalization. These are the people, who master and know naturalization, that is about the nature, of man and God, and about the God, who control, and masters the universe. These are not where our thoughts come from, but where His does. This aspect, of control, and manipulation can only be mastered by Christ, the Lord. This is how I write this book. This is through naturalization, and the ways, that power, and influence, reconstruct and recreate a world, that is full of knowledge and self-control. This is about the absolute power. This is reconstructed through the sins of the earth. These are recollected, as the order of the universe, that is wholly about the ways that the order, is about the knowledge and ability, and about wholly, about how the universe, and the order, is directed toward progress, and the abilities and motivations, come from the mastering of these, and the ways that power, and authority come from this, is about the context and contention span, of reality, with the thoughts and thinking, about and for the essences, of the Creation. The creator created the creation, and creator to be after the image, of Him, or "Us". These are "Ourselves" which is a meaning beyond the contention span, and realities, that we construct and create, in our own image, except after His, which is knowledge, according to this theory. This is about how He did not like, anything that was not in His Image, but with the world's view of knowledge, and the power, of control and authority, that is about the authority, that is about the authority complex, and what this means to a preppy kid like me, with absolute divine power also, in addition to immaculate knowledge also, which is fashioned after the image of what I have also seen. This means that I am also God, or (part of him), which is a meaning beyond reason, meaning that we have absolute control, over nothing, but knowing God. This is

about the ways that the God, of the universe, are about the contextualization, of the contestation, of knowledge, that is also about the direction, and influence, of power, and the authority of power, and the authority influence, that is about the God genetics, that are not about the doctor and patient analogy, but about the direction that we influence, that is about the God, of the universe, and what He is about, that is about the context, that is about the Biblical Interpretation, of Jesus Christ. He was all of this. This is about the context, that is about the ways that the content, of the context, of the nature, of man, is about the context, and knowledge of God, and man, and about the mankind, that was about the man, and his God. This is about His maker, and how the creation of his was to be simple, and knowing, and perfection in ways of love, and good, and mastery. These are about the conflict of interest, that are about the knowledge of man, and how man and God, are about and are consisting of the issues, of the divine influence, and how the divine even helps, and guides, the maker to His creation. This was after the fall of man. These are about the divine instructions, that help and concern, ourselves, with the creation, that is about the creation and context, of knowledge, that is about the context, that is about knowing, that is about the knowledge, that is about the contextual analysis, that is not in this book, but in the manual for naturalization. This would be the art. Also, the science would occupy space, in this venture, of accomplishment. Also, it would be science. But the message is that we are free, and naturalization is about the guidance of nothing, but the realms that you have experienced and how you get into them. This is about the postmodern world, and the revelations that it has promised and concerned with us, to be helping, and not hurting, that is an example of Godliness and Godlessness, that is and are

about control. To control ourselves, we must have naturalization, that we understand and comprehend, that are about the science and art. This is of becoming what we want to be. These are the comparative analysis of the nature, of man, and his art. These are all having to do with time, and place, and Godliness, or Godlessness. This is created after evolution. This is why this is true, that naturalization is in evolution too, just because the creator liked this, through myself. The opinion of a man is more valuable than the research of a doctor, but the patient never figures this out, which is a mystery. This is to God, and man, and the approach that he learns from, and teaches others, to become, and be with, that is about the control, and authority, of man, and where he goes, which is into the other world, of creation and manufacturing, of thought, and idea and opinion that is wholly his, with ideas, and constructions, that are about, and for him. These are not the "gay science" but the other way. This is about God. This is about the God of the world that is about the way the world is about the control and authorship, that is about the substances and essences of realities, that we learn to earn and control. This is how naturalization is accepted with God, and man. This is about how the grace of God, and the gift of Jesus Christ, is through the gift of the Lord, God, who created us. This makes sense if we are in His image, or creation. This is about creating ourselves whole with Him. This is just a check away of naturalization, that is about the nature of man, and His God. This is about who reaches Him. These are about curses of naturalization, on man, because of solely evolutionary judgmental thinking, that makes no sense, so is not true. The doctor and patient analogy is about the control of the judging and control of the complexities of man, and how he adapts. This is about the control of the complex and

authorize outcomes, of man and mankind, and nature, and man, and his issues, that are controlled by naturalization, and its excuses. These are that it created God, and is involved in power, and authority, and the complex individual that is issues and complex related that is about wholly issues that are powerful and complex and concerning for others. This is pure false. This is unless the "others" are Christians. These are not to be related with God. This is in the sense, of disbelief. God, must be believed in as true, and whole, for nature to take its course, and create and destroy what is embedded in our hearts. This is our destination. The control and authority of the God,, of the world, that is about the embedded philosophy, in our lives, which is about the control and authority of God, which is in the thought and thought-control of our reality, that is made up to be issues, of relevance, and authority. These are about the naturalization.

These naturalization impulses of the mind, are about the conflict and interest of the gain and loss of society and the absolute and authority of the individual, with God, and with the Image of God. These are just an explanation of naturalization, and how this naturalization works, through God, the Father, and Jesus Christ, His Son. But, the response to God, is about the essence and substance, of God, and how He works. No work could ever be worthy of God. These are about, God, and about how He is about the capable and able, creation of the essences and substances, of the world and is about the Divine power that is of and are for and about the confliction of interests that are about the abilities and outcomes. The way and the truth and the life is through Jesus Christ. These are about the control, and authority, of individuals and the way that individuals live their lives and obey things and learn

about the world. This is always through these five 1. Different Elements of Designation in Thought, 2. The Thought of Deformation, if Learned, and 3. The thought that has gained an explained outcome, and 4. God. The naturalization is able and willing to be seen through the gain and loss, of individuals, and gain, and loss, are through the outcomes and willing and capable strengths and weaknesses, that are controlled by the God, of the world, and willingness, to controlling and gaining contemplation, and the weaknesses. These are through these four things. These make up a good plan. These plan oriented, outcomes are involved in the aspects of willingness and gaining to the world, of gain and loss. We all must learn through success and failure, which is the rule of naturalization. These are agents, of philosophy, that must earn and gain, acceptance through the outcome of the world, through the aspects of philosophy, and the "thing in itself" which is not true, but also an outcome that is generated, through thought, and thinking, and absolute. The way that "thinking in itself" is not true, is because the "thing in itself" cannot be created, unless it is equality. Yet, the scientist that created this, is not a democracy citizen of the free world, but a idealist, of common sense. Yet, this does not make sense either, to man. This does not make sense to God. Yet, it makes sense to the philosopher who wrote it, Heidegger. He is the one who knew of the "thing in itself" and the outcome of naturalization with this, is impossible, but also possible, except with God. This is also about the philosophy of Nietzsche, and how he cannot impress, this. This is through Godlessness, that is about absolute power, that is about the reflexology of nature, that is also man, and that is also mankind, that is about the "as is". This philosophy is also understood. This is in the nature of the equation of anything, possible,

with anything, that is possible with naturalization, and how the nature and man, are about the naturalization, of man, and his possibilities, that are except, and besides, the natural. Man can have a natural state with God, in the "Dassein" and how it operates, is through the impression, and impact, on others. This is through the philosophical, impression, on the self. This is what is immortal. The other in "Dassein" is about the conflict, of God, and man, and how we must be able to control, ourselves, within the context, of our journey, through time and space, and how we must relate to everything, through the "thing in itself" and of others, and about how they must control and react, to impossible, and outlandish odds. These are all created by the "thing in itself". This is therefore a moot theory. But, it is not. This is how I compare God with the thing.

The thing is about the that that is about the that that is about the that that is about the that that is about the that that is about philosophical enterprises. These are whole new realms.

The thing in itself, in philosophy, is God. These are about different languages, and how you can use them through the context, and conflict and interest, of deformation. This is about the church. These are about how the churches are about God, and are about fear factories. These are right and holy. This is about the conclusion of "Dasein" and how it is right and holy if you think of it this way, through the eyes of the illusion, that is simple and perfect, yet also right and wrong, and about the conclusion in itself, that is about the gain and loss, of itself also. Yet, through "Dasein", there is a hidden meaning, beyond the words and comprehension, of the loss and gain, of itself, that is beyond reason and comprehension, that is also not. This is about God, and the reason why He exists. This is

about the theory of naturalization, and about how the "Dasein" of nature and man combine. This is to create something else. These are about the ways the followers of Jesus Christ, are about the combined interest of God and Jesus, and how they are about and for the followers themselves. This is due to God and man, and nature that explains itself through anything it wants, due to sin nature. These are nature of God and man, through "Dasein" that is natural also. This is with God and nature. The reason why Dasein makes sense is about and due to the cause of naturalization, which is also about Dasein and about how the conflict and interest of this is due to conflict and interest in itself. These are and would be the part of the nature, that is a part and natural philosophy of finally the self. This is what wrote naturalization. These are what these parts of naturalization, are about, and of, that are of and about the conflict of and interest of the man. This is the true writer of naturalization. These are through the aspects of naturalization. These are about and for the leaders. These are about and for the followers. These are about and for God. These are about the resolution of the grand façade. This is about the plaything and its every desire. These are about the holiness and how we must have and keep it. In the order of Dasein there is a message that makes sense. These are that the message is holiness and how we must have and keep it, in itself. This means that the order, is about holiness, and how we must have and keep it, in and of itself, which is about God. This is about time and senses. These are about the nature and man, and how God, and nature, and man, are about the thinking, that are and is of and about, the cherished outcome, and how we must achieve it. This is about the Dasien, and how one man wrote this. This is about the self also. The dilemma of the self, is through the own di-

lemma, of mortal man. These are his followers. These are the followers of books. These are the followers, of Jesus Christ. He is not the word, but the Holy Word. He is about the Bible, that is about and of the followers. If Jesus Christ didn't have followers, He would be acceptance to the people, below Him, without being this, but being a true follower of all others. He was God, though so, Dasein, does not believe in Him. It believes in its own theory.

The way that man is about fact, is about, how the fact, is about the fact that is about the way. This is about the truth. This is about the life. This is about the light. This is what the control is about, in the doctor and patient allegory. This is about the life, and about how, we have the life, that is about the living life that is about the God, that is about the life, of the living, and the life, that we have accepted, through the God, of the universe, and His followers. These are the people, who wrote the Bible. These Bible stories, are about the generation, and the followers, of the generation, of the people. These are about time, and how Jesus Christ, created time. The time, is about, the progress, that is about the youth, that is about the design, that is about the outcome, that is about the process, and ability, of God, and Jesus Christ. This is about the ultimate parable. This is about the process of thought and thinking, that is about the message, and authority, that is about the God, we love, and about the time that we have, which is ours, which is for Jesus. These are about the Accolades, of praise, and reason, for such a humble theory. This is about the holiness of God, and about how the God, of the world, is about the conclusion, that is about the time, that is about the essences, that is about the grade we make in school, the position we have in church, and the goings that we have in the government, and how we make

individualism, from these, These are for our own "Dasein", and conflict and interest, that's is about the "Dasein" and about how we must choose, and be capable, and willing, of the acceptance, and willingness, through the choices we make. The choice I made in writing this book, is through the Jesus Christ. He is the maker of earth. He is the maker of heaven. He is the one true God. These are about and are of and for, the able bodied and willing, people, who make us happy. These are all about God. There is a contest to see who is the best God. He is the one true God. He is Jesus Christ, and He is the creator of the universe too, in three persons, that are the trinity, of God. This is about God, and how He must be controlled by Him, and about Himself, that is about the way that Christ is about God, is about the God of the world and how we are interested in Him. These are about the God. This is about the way that God is about the control and adaptation of the world's dilemma which is the dilemma of the self. This is about the word and how God is about the word, that is about the followers of his, and how they are made up of one and another person, that is about and of one and another follower, that is about the one and another way that naturalization explains itself. This is about the true and truth God that is about Christ, and Christianity, and about the afterlife. This is all explained through naturalization in a real way or I am a liar. These are about the one who lost the war, that is about the one who won the war, that is about the one who lost the war that is about the one who won the war. These are about questions and answers, that are about the conflict and interest, of one man, and one God, and about God and man, and how he is saved. If this is wrong, in any way, I am a true hypocrite. And, also if these are the wrong words, after you learn them, you are not a liar, but a smart person,

who knows that this is absolutism. This is about the way the doctor, patient, allegory is from and about the trial of Jesus Christ. This is about the subjectivism, of the approach, of absolutism, and how the doctor, and the patient, were crucified, with each other, like the God and the disciple, are also this same way. The doctor must crucify the patient, sometimes, in the words of hypocrisy, because of someone else. This is the biggest lie ever. This is about the ways that man, and nature, and mankind, are about the essences, and substances, of man, and his following. This is a matter of how true, you can see, the world. These are about the ways that man, and his followers, are not crucified ever, unless it is by the one you knew did it. This is in God's name. The explanation behind this, is about the essence, and substances, that are and of about, conflict with reason. Yet, the only one who crucified Christ, was Judas. The Word was according to this. Every other person, practices falsehood when they do this. And, the first and last person in my life, to do this, was Charles Darwin. He is the major in the crucifixion of Jesus Christ. He is about crucifying Christ, with his own deeds, and words, and actions, through literature. But, I never knew where he came from. This is in the Bible. This is about the way, that the world, is about the details, of Christ, and how He can communicate. Jesus would say, that the world, is about the world, that is about the world, through Jesus Christ, and the Word of God. The theory of nature and man is about the able naturalization and about the absolute truth of the Word of God, that is living and moving. This is through soul, body, mind, and spirit. This is about the way the world is about the way the world is about the way the world is about the way the world is about the way the world is about the way the world is about the movement of God, and His followers. This is

about the complex and complete, ways of nature, and man, and about his followers, and His disciples. These are about the followers, and the following, that is about the following, of Jesus, and His own maker. These are about the productive and outcomes that are simple and complete and have the complete and simple ways of completing the ways that the world are about the God that is real and about how the way that this is about the ways that the world is about the complete and the complex ways that are about the way that is about the ways that the world is about the nature of God, that is about the way the way the world is about the way the world is about the way the world is about the salvation that is of and about the complete ways that a code is about a universal substance that acquires knowledge. There is no code for naturalization, but there is a language. There is no code in language, but a complex code of matrix design. These is about the code of the universe. These is about words. The math of the world, is through symbolism. The science of the world, is through math and words. And, the reflexology is about the ways that the dynamics of the world, is about the systematic functioning that is about and are for the matrix designed to perform functions, that are about the knowledge, that is about the ways that are about the God of the world and about His knowledge. This is about new and newer than old, complete and new ways of justification. These are about the lies in the Bible. These are not true, because the Bible is Holy. It is about the complex ways that the new is about the old, and the new, is about the way that the Bible, is about the old, and new ways that are about complexities. These are about the ways that are about the complex, and complete, ways that are about the ways that are about naturalization, and the ways that are about the problem and about the solution. This is just

cause and effect, that is about naturalization, that is about willing and free, nature and man, and how we relate to everything. There is a lie that is about the Bible, and this is about Satan. He is only an angel of light, and the Bible says that he is an angel of darkness. Yet, the Bible can be translated any which way you want. This is not the language of deformation, that is about the way that the world, is about the new and new about the way that new is about our country. This is about the way that the world is about the way the natural way that the world is about the content, and context, of the basic instructions that are for the world. This is what is about the way the world is about the way the world is about the content and complex situations that are about the situation that is about the way the world is about the way the world is about the way the world is about the way that complex situations the world has seen is about the way the world is about the way nature and man, are about each other. No one is guilty, but everyone is innocent, and the way that man, is innocent, is about the way, that is about the way the world is about the holy spirit and how the holy spirit is about the complete and operating ways that man is about the way that man is about God is about the complete and total ways of living, that is about the way the world is not about the way the world is about the way the world is about the way the world is about the way the world is about the way the world is about the way the world is about God and Christian theology that is about the complete and total way that is about the complete and total way. This is Satan's, theory of naturalization, and how He came into being, and how he exists. This is not a manual of lies, that is about the truth of Jesus, but about the truth of the naturalization that is about the ways that the world is about the naming of words, and how the world is about the way

the world is about the complete, and total, ways that are about the direction, of lies, and how Satan must be King of them, because of the God of the world, and how God must control, the world. This is about the plan, that is about the complete, and total ways, that man has controlled, the life of the world, and how followers of the God, of the world, are about the complete and total, ways that are about the way the world is about the way that the world is about the way the world is about the way the world is about the way the world is the oyster and we are the pearl, and not the other way around. We are the precious thing, and the clam is our outer covering, meaning that we are the substance and the essence, and the life of ours is about the protection and covering of the shell. These are about the dynamics, and how we are interrelated with the system, and the designed way of the outcome. These are about the performance that is about our God's. These are about the ways, that man and nature, are about the complete, and complex, ways that are about the ways that are about the way, the way that are about the timely, and essential, ways that the mask of Hitler, and other, bad people, were not influenced by Satan, but by pure intentions, for war. Satan, is about the light of the world, and about the way the complete and total way that man is about nature, is about the complete, and complex, ways that nature, and man can and will exist. These ways that man and can be exist type of thinking, comes from the issues and relativity of the reflexology or termed word for the issues, is about the dynamics, of the interwoven, math of the science, and the words that are the theory, of dynamics, and math intentions that are. These are interwoven into the very fabric of existence, that is ours, and yours, and us and them. These are the issues, of how we see, us and them, and about the outcome. These are the outcomes that are

about the senses, that are about the way the world, is about the outer shell, that is about the way God, is about the ways, that God is sane, and influential. These are the very fabric of the core, of the analogy of the doctor and the patient, ratio. These are about who believes, in this, and who comprehends this equation, in a basis. These are about the complete, and the complex, situational, issues of thought and thinking, that are about the whole, issues, of relativity, and the ways that the relativity, is about the dynamics, of learning and the speakeasy language of deformation. This is about deformation, and how the love of the language is about the real and the fake awards and recognitions, that I expect from writing this, unless people call me a liar. These are the complete, and complex, the ways that we seeing this and hearing this and feeling this with the senses, is about the drive for success, and the failure of the book, that is about the ways that the dynamics, of the interwoven fabric, is about the design and influence of the fabric, that is of the interwoven, fabric, that is of and about the complete, and total, ways of understanding, and organizing. This is about the reflexology, and the meaning of the all of all of this. This is about the way that the people, that are about the people, are about the ways that are about the complete, and complex, ways that man is about nature, that is about man, that is about the nature, that is about the mankind. These are about the complete and complex ways of understanding that are about the ways that are about the control, of the population. This is through Christian naturalization. This is what the world, is about that is about the nature, of God and man, and about the man and God, that are about the ways that man, and nature are about God is about the way that nature and God is about the conclusion, and the result of man. This is about the ways that God is about the fall of

man, and about the way that man is about the way the man is about the issues that is about the way the religion is about the occupation that is about the way that religion is about the way that is about the way that religion that is about the way that the world is about the way that the world is about the way the world is about the way the world is about the way the world is about the way the world is about the way the world is about the way the world is about the way the world is about the way the world is about the way the world is about the way the world is about the way the world is about the way the world is about the way the world is about the way the world is about the God of the world and how He is about the way the world is about the disassociation of the way the world is about the theories that don't make sense and what the theories that make sense are about the way the theories that are about the way the world is about the way the world is about the unique aspects to the world that are about the conflict of interest. This is about the way the world is about the way the world is from and about the way that the way of the world is about the way the world is about the God is about the way the world is about the way the world is about horror and abuse, and how the people does not stand for this. This is in any context. This is about the way the world is about the way the world is about the way the world is about the way the world is about the violence of the world and how it must stop. This is about the way the world is about the way the world is about the way the world is about the conflict of interest and how the issues of the relevance of God is about the way the world is about God and Jesus. This is about God and His issues, that are about the man, that is about the God, that is about the man, that is about the God, that is about the way the world is about the way the world is about the way the world is about the way the

world is about the complex and complete world and what is in it. This is about naturalization that is about the way that the nature that is around and always about us. This is about the way the world is about the way that the world is about the ways that God and Jesus Christ are about the ways that God and Jesus Christ are about the naturalization. This is about the naturalization that is about the way that the world is about the way that naturalization is about the way that naturalization is about the process of becoming. This is about the way that the world is about the context and content that it must understand as a whole. This is about the dynamics and what dynamics are about. These are about the complete and complex situations that are about us, with a math and physics related equation. This is about the way that is about the way the world is about the way the God of the world is about the way the world is about the way that the world is about the God of the world and about how the world is about the way the world is about the creation and explanation of the world. This is about the way the world is about the way the world is about the ways that the world is about the content and context, that is about the evolution and how this is not true, but very fake, like a reality that is never about belief. This is only about trust in your own evil. This is about the things that make us who we are. These are about the way that makes us who we were already, and already again. These are about the things, that make us who we are and are about, and for. These are the naturalization aspects of man. These are the aspects, of the fallen man. These are the aspects of nature, and man, and how nature and man, survive. If one takes life away, he has life in the first place, until. This is until he guarantees himself an escape from the society. This is for Jesus Christ. This is about the way the nonchalant ways of jus-

tice, and the ability of justice, helps us out, in Christ, because they are laws. Laws were a creation of Jesus Christ. This is why he is a prophet. He knows the plan of His followers, to the end. And, I am one of them. These are the followers that are about the nature, that is about man, that is about the concern, of man, that is about the ways that are about the ways that are about the ways that are about the ways, that are about the ways, that are about the ways, that are about the ways, that are. These are about the conflict and interest of modern man, that is about the speakeasy philosophy, of God and man. He is about the sense that is about the senses, that is about the senses, that is about the senses. This is about the reflexology of opinions and how we are about them. The way that we are about the opinions are about the way that the opinions are about the God of opinions and how He shapes them from us, in total factuality, meaning that He is God, and what I am writing is about Him. They are also about me. These are about the ways that man are about the ways that man are about themselves that are about the way man are about themselves. This is about the way the philosophy of the learned are about the ways that man is about the process of factual realities that are about the senses, that are about the ways that are about the way the world is about the conclusion that is about the sense of creation that is obvious. This is about the sense and senses, that we all learn with, that becomes with faith, and belief, that is about the obvious. This is about the way the world is about the way the world is about the faith and works of the world and how works are about faith, or they are wrong. This is always a misleading philosophy. These are about the senses, that are about the senses, that are about the senses, that are correct, with knowledge, until the aspects of man are combinations to create the world.

These are about the ways that man is about God, that is about man, that is about God, that is about the man who is about Godliness, and created outcomes of Godliness that are about predestination. Why do we even think, if we are about predestination? This means that there is a real plan, that is about the Godly, that is "already there, and we cannot change it". If God knows all of our thoughts why don't we act in the way we should always, and be forgiven for our sins, always. These are not strange or wrong points, but points that makes sense, to a higher level, of God, that is about Jesus Christ, and His calling. This is how He is about the control, that is about the authority that is from and about God. These are creations that are about God, and how He created the world. If He can create the world, then He can predestine us, which is similar. This is about the ways that man is about God, and is about the ways that are about the sense that is about the control, and orientation of God which is about the conflict, of the interest, of man and God, and how it works with Jesus Christ. This is about the way Jesus Christ, works with the essence and substance, that is about Him, and God. He is the savior and God is the creator. He is the one who creates things, and Jesus is the one who knowledge wise, has all. This is about the way that the world is about the God of the world is about the real Revelations, and how Christ will return for a new Garden of Eden. These are the evenings and mornings, of the creation of the world. We still have these because of the creation of the world. This is about the way that the world is about a good deal, that is about our salvation. These are about the ways that the world, is about the goodness, that is about the sense that is about the sense that are about the control of the world. This is already created and already saved after Jesus came. He recreated the world, to be His, as well as God's,

to save all of us, with the most importance of God, who is in control, of the world, that is about the ways the world is about the God of the world with His schedule. He plans out the world. This is his plan. This is about the ways that the plan is about the complex and complete ways that God and man can behave and control their behavior, or we would symbolically be monkeys, for Darwin, and His creation theory, that rules the world of misbelieving things, This is about the way that man and his nature are involved in the creation of religion and belief. We must have a physics theory that is about Christ, to know ourselves differently. The fact, that the order of the thought of man, is based on naturalization is completely true, and completely relevant to the world around us, in the manner of thinking which is and are about the Bible. This is about the way the dinosaur is about the egg, that we either the "or the egg" symbolism, meaning that the order, of the world is about the senses that are about the learning that is about the creation in the doctor patient analogy. This is a meaning saying that the world and the order, within, is actually controlled. This is not our control, but our environment, that is about the ways, that is about the problem of the "chicken or the egg, which was what came first", is about the evolutionary theory, and the progresses of the way that the dinosaur was extinct. Was this before or after the factual existence. Was the chicken or the egg created first? This is a philosophical statement, that does not explain anything. Yet, naturalization explains how God created the universe in basic terms, through a physics equation. The order of the factual existences are based on the God of the world and how He created it. But, then, why did we fall. This is due to knowledge. This is about knowledge and about how knowledge works, that is about a chicken laying an egg. This is about the functioning and

functions of the way that God is about the functioning also. This is about the context, and content of naturalization, and about how the naturalization of the world is about the challenge, and the exemplary method, that is about the progress, and achievement, of the challenge, and how we handle it. This is about the way the world is about the conception, of the world that is about the content, and context, of the naturalization of man, and about how the man, is about the man, that is about the man, that is about the man, that is about the man, that is about the way of man, that is about progression, and contextualization, of the essence, and substance, that is and are about the way that man, is about the way man is about the liars, the cheaters, and the thieves, that are about the way the man, is about God, and what is that, that is about God. This is about the way the man, is about the way the man, is about naturalization, that is about mankind, that is about the man, that is about the man, that is about the mankind, that is about the content, and the way the contextualization, is about the construction, that is for the way the world is run, with content, and contextualization, that is about the lie, that is about the stealing, that is about the cheating, that is of and about always, the redemption of Jesus Christ, and how He can forgive you. This is even through naturalization, if you believe in this, power that is about the ways that man, is about God, that is about Godlessness, and Godliness, that is about the power, and authority, that is about the essences, and substance, that is for and of God. This is how the story tale ends. This is about the way the world, that is about the ways that are about the ways, that are about God, that is and are about God, that is and are about the God, of the world, that is about the way the world is about the God, of the world, and about how.   This is about the way that God is about

the ways that man is about God and the way that naturalization, is about the content and context, of natural things, and how naturalization happened. This is through the perspective. This is of our own world, that is about and of the essences and qualities that the world has, and has been bought with, including with power. This is about the symbolism about how we own everything, like owning the whole world, including the way we think, influenced by the money system, that is a power system. This is about the way that God, is about essences, that are about the Godliness, that is about the Godlessness, that is about the creation, that is about God and Godliness, and the construction of man. These are about and in God. This is about God. This is about how God, includes Godliness, through the system. This is what I have designed. This is a system of thought, through power of explanation, and guidance of power, and explanations, that are about the essence, and context, of nature, and man, and nature, and man, and nature, and man, and nature, and man, and about how God sees this, through the entire perspective, as a whole. This is about the man that is about the man, that is about the man, that is about the man, that is about man, that is about man, that is about mankind, that is a whole bunch of lies, that we should not believe in, and according to God, ever give. This explains the system. If we know something, it hurts us, as well as helps us, besides the way we act, that is about our belief system. This is naturalization. These are in, naturalization, that is about the ways that naturalization, is about the progress, that is about the ways that the progress, is about the way the essence, is about the essences, that is about the substances, that is about the way that the world is about the context, and how the content, is and are about Kay Bailey Hutchison, and how politics are not always about sex, but

about the control of the world through man. He is the one responsible for our essence, and our freedoms, and if he takes this away, it will be wrong, of him. The "man" is someone who knows things, about others, according to my learning, about others. He is the dictating one who knows all of the world's truth, and lies, and conspiracies, that are of and about, the future, truths and lies, and then only his opinion. The girls are the ones who know about themselves, and the way, that man has self-control, and resistance, toward them. This is if they are in Christ.

This is about the way that the world is about the way the world is about the way the world is about the way the world is also about the complex, and complete, unknown, intelligence, that are about given, and responsible, that is about the complete and complex, truths that are about the way, that man, is about the mankind, that is about the option, of choice. This is about the sex. This is also about superstition and how this can be this way due to strange beliefs. But, according to me, sex, is not a choice, but it is natural, with natural choices, that are about the options, and choices, that are about opportunity, and the helps, that the world gets. This is sometimes corrupt. There is a strange resemblance between sexes, and that what sexes have always been about, are about the opposite sex, thought of as male, and female, Why do we mess with God, in our creation? This is the question.

These are the days, of the choice of free expression, and the ways that mankind, has dealt with free choice, is about the way that Satan, has tricked us, into believing in a strange world. This is the world of monkeys, and men. Monkeys do not have anything to do with mankind. This is why Darwin, is a cheater, thief, and liar. This is my only

conclusion. This is because Satan is famous. The way that the nature of man, is about the ways that man, and is about the mankind, making of Satan, through belief only. Yes, we can defeat Satan by just not believing in Him.
This is about the way the world is in the Bible, with Satan, and has many different facets that are about the complete, and complex, ways that man, and mankind, are about the context, that is about our existential ways, that are about man, and mankind, and God. These are about how God is superior to Satan.

1. He is God.
2. He is the creator of Satan.
3. He is smarter than Satan.
4. He is the most influential figure in modern history.
5. He knows all the history of the world.
6. He created all of the world.
7. If you believe, in God, then you can be part of Him forever.
8. He can save you.
9. He can explain anything to anyone, through prayer.
10. He is a living, breathing, being.
11. He is in Heaven.
12. He knows all of this.
13. He is the author of the Book of Life, through His Son.

Satan is not these things. But, to be scared, Satan is a false God. He is a death angel. He is the vagabond of every person. He explains this to himself daily. To not be scared, one does not have to believe in Satan, and he is usually saved.

God is the only one who believes in Himself. This is the way the world does things, and operates. This is about the way, that God, is about the belief in the world, and how it is for Him. This is about the way that the world is inhabited by crazy and sane people. They do not disagree with each other. This is about the discourse of God. Jesus Christ, is in all people, who believe in Him. He is the one and only God who is a true creator of the universe, besides His Father. He is the One who created the whole world.

The ways that mankind, and man, is about mankind, is about man, and about how man, is about the mankind, that is about man, that is about the man, that is about the mankind, that is about man, that is about mankind. This is all about the way we are controlled by God.

The way the world is about the way the world, is about the grand illusion, is about the ways we look at the world. This is through the eyes of the perspective, of mankind, and man, and about how the man, and mankind, that is about the man, and that is about the mankind. These are about man. These are the rules and principles of mankind, that guide and control us. This is all of God. This is about the ways that God, is and are, about Christ, that is about the guardian, that is about the providence, that is about the expectations, that are about the control, and guide, of the universe, through God, and about how the Christian, Jesus Christ, is about the control, of God, and how about the control, of God, is about the control of Jesus Christ. This is about Jesus Christ. He is the God who is about the context, that is about the content, that is about the controlling aspects of God. God is about Jesus Christ. Jesus Christ is about the control that is about God, that is about Jesus Christ, that is about God and Jesus Christ, that

is about the control, of God, and about the man that is created by God, and made with His voice, and which is about the creation of God, that is in Christ Jesus. He is the one who created us, to save, us. This is predestination. The world in a nutshell, is about the control, and authority, of the essence, of mankind, and about how mankind, is in a system. This is about the beliefs he has. If we do not believe in Jesus, we do not become saved, and go to hell. This is always written in the Bible. This is what the Bible is about that is about the Bible. This is what the Biblical translation of the Bible, is about and for. This is to win, in the world, that is about and in the world, that is in religion. The fact, that is about the way of control in the universe, is about the way that the world is about the control that is about the way the control is about the direction of the world that is about the real Jesus Christ. This is about the way the world is about the way the world is about the way the world is about the clubs, and formations, in our life, that are about the ways that the judgment of Christ is about the way the world is about the way the direction is positive, when you have these things, that are there for free, that are always around, and about the social consciousness, that is about our world. This is not a concern for the day that we have. This is a signal of promise. This is about the life that is ours and how we must communicate with the world that is around us. This is about the world and how the world, is about the truth, that covers us, with the essence and substance. This is about life. This is about the way that the shape of the world is about the way the world is about the way the world is about the way the world is about the concern, that is about others. This is unless you are in a club, or place where you put your life in. This is also, the Christian religion. This is of the society, that is of and for, the best. This is what the

best is. This is about the rule and progress that is about our world, that is about the way s that the world is about the ways, that the world is about the great and simple, compromise, that are about the class. This is about the way the world is about the way the world is about the concern, that is about positive, and negative approaches. These are about Jesus Christ. These are what the world is about, that is about the world, and how we communicate with what the world wants, and what the world gets. This is about the way the world is about the God of the world and about how the world is about the way the world is about the way the world is about the way the world wise, ways, are about the ways that the world is about the way the world is about the way the world is about the gay, and the straight, and the sexuality. This is also plays a part in the real world. This is about words. These are what eventually hurt us, and harm us. These are what we can learn, as the Bible, and how the Bible, works, is through the Biblical interpretation of the Word of God, and how about the Word of God, is about the way the Word of God, is about the way the world is about the way the world is about, the faith, of the world, and about, what the world is about, is about the nature, of the world, and about the way the world is about the God, of the world, and how He still controls this. This is about the ways the world is about the way the world is about the way the world, is about the science, that is about the way, the world, is about the science, that controls the way, the way looks at us, in science, and methods of thought that are about the ways, that the method of thought, communicates with us, that is unknown. This is unless we understand deformation. This is about a meaning, meaning, that the order of thought, is about blame, and insane notions, of thought, that are about the process, that is about what thought is about, is

about the thought itself, and about how the way, of the world, is about nonconformity, and about the way, the nonconformity, is about the social club. This is about the way the social club, is about the science, and the way the sociality, is about the way the world is about the nonconformity, that is about the way the world is about the way the world is about the way. This is of the world. The ways that I have seen, and interpret the world as, is about my naturalization, and how the world looks to me. This is as sane, and insane, and bad and good. These are the constructions that I am the most familiar, with, that makes sense. This is about the way s the world are about the way that the world is about the thought disorder of the universe, and how the thought disorder, is about the way the world, is about the ways the world is about the sane, and insane, notions that are existence, and about the existence, that is and are about, the direction, that is about and are, for the about. These are what are about, the essences and substances, that are about the direction, that is about the following, of the thought disorder, and about how the world is about the direction, that we follow, and about and become, what is about the generation of thought, and how we generate it. This is through the justice system. This is through the social clubs. This is through the money system. This is through the governmental monitoring, and about how the world, is about the way the world is about the social, and the aspects of Christianity, and how the Christ of the world, is about and above, and beyond the point of repose, that is about the direction that we follow, and the direction, that we all go toward, and against, that are about, and for and to, and from, the world, and about the world, and about the ways that the world, is about the way the world is about the way the world is about the way the world is about the

way the world is about the way the world is not about Christianity, but is about the insane, sometimes, and how we feel, is about this, which is a burden, to society, that is about the direction. Some people, and not focused on the result. This is about the clinically insane. This is about how the clinical doctors, are also this. This is about a broad opinion, that is about the direction, we go, and first and last, becomes what we know about. This is of and about the way that the world is about the way that the world is about the way the world is about the way the world is about the way the world is about disorder. This is about thought, and how it are not communicated, together, but are separate, by a huge degree. This is about the thoughts, that are about the control, that are about the senses, that are about the control, of the order, of thoughts, of thinking, that are about the disorder, and how about the ways that the disorder, is about, and are about, the thought disorder, is about the ways the world are about the ways that are about the science, that are about the controls, that are about the controls, that are about the controls, that are not about the controls, that are about the client that is about the relationship, that is and are about lying. This is not the right way, to live life, to the fullest. This is not the good way, that people, live, and is about the way, the controls are about the controls, are about the controls, that are right or wrong in essence. These are about the controls that are about the controls, of the way the world is about the essence, that is about the fake, and the real, and about how the life, is about the control, that is about the way. This is about the way the world we live in, is about the way, the world is about the way, the God, is about, the science, that is about the way, the world is about the world, is about the way, the world is about the way the world, is about the judgment of Jesus.

If there, is no judgment, then we are not insane, or hypocritical, or self-righteous, and there is about and from, the essence that is about the client of the essence, that is about, the right, and wrong, ways, that stack up, and make a difference. These are about the control, and management, of the living life that is ours, and the jealous and envious people, who are not sane, but real big liars, or cheaters, or thieves, that make sense. This is to nothing. There is a secret of the life of anybody's that is about, the control and manipulation, of God, and how He must judge us. This is in the image, of God, that is about the image, of Jesus Christ. This is about the policies, of justice, that are about the ways that the world is about the ways that the justice, of anybody, is about the science, and policy of the world. This is about and through the justice, system, that is about the science, and clientele, of the body, that is about the essence, that is about the substance, that is about the science, and the ways that man has created, and destroyed, life. These is within the world. This is within the world, and within the origin, of the world, that is about scapegoats, and modernity that man is part of, and in control of, and about the research, of man, that is about the science, that is about the prodigy, that is about the science, that is about the ways, that is about the science, that is about the science, that is about the science, that is not about the science, that is about the fair, and the game, that does not exist, but does exist, in the patterns of thought, that exists in us. These are of the science, that is about the conclusion, that is about the way, that the disorder, that is about the science, that is about the science, that is about the science, that is about the natural, way, of acting, that is about the are about the relationships, These are what we have in life, and in living, and in science. These are about the science, that is about the control, that is about the sci-

ence, and that is about the essence, that are about the science, that are about the science, that are about the control, that are about the science, that is about family, and how they work together. This is about the control, that is of and are about, the senses, that cannot be controlled, or manipulated, that are about the science, that are about the control, that are about the science, that are about the Royal family, that is about corruption. This is about the way that the world is about the ways, that the world is about the corruption, that is not its, but Satan's. These are about controlling strategies, that are, about the corrupt, and gentle, Jesus Christ, and how these have nothing to do with each other. This is about how people, beat, each other up, and senses adjust to the world, that is within and without, the details, of this science, and about each, and every understanding, that is about the control, that is about the conclusion, that is about that Jesus Christ controls our world. There is nothing that can separate us from the faith of God, that is in the world, through Jesus Christ, our Lord. This Is about the way that the world is about the life, of the world, that is about Jesus Christ, that most people don't see. This is obvious. Yet, the royal family, is about the corruption, that isn't about them. This is about Satan. These are about the way the words, are about the individualism, that is about the control, that is about the process, that is about the thought, that is about the thinking, that is about the thought, and about the thinking, that is about the thought and is about the thinking, that is about the thought, and thinking, that is about the thought, and thinking, that is about the thought, and thinking, that is about the thought. This is about the thinking. This is about the way, that the world, is about and for, the reactionary, impulses, that are about the ways, that man is about the God, that is about the think-

ing. These are about the thoughts. These are all about the thinking, that is about the thoughts, that is about the thinking, that is about the thoughts, that is about the thinking, that is about the thought, that is about the thinking, that is about the thought, that is about the thinking, that is about the thought, These are about the thinking. These are about the thoughts and thinking. These are about the processes of thought that do not become judged, but exist, independent, of the notion, of being, and situational, analysis, and completion, and concern, and about the science, that is about and are about, the science, This is of man. These are about and from, the essences, that are from and about the sciences that are from, the science that is from, the essence, and from, the problem, that is from the essence. These are also from the substance. This is about the essence, that is about the substance, that is about the problem, of science, that is about and for, the essence, that is about and from, the essence, that is about and from, the essences, that are from, and about the ways that are from and about the way that are about the ways, that are from, and about, the science,. This is about and from, the science, that is about and from, the sciences, that are about and from, the essences, that are about and from, the essences, that are about and for, the essences, that are for and from, the essence, that is about and from, the essence, that is about and from, the essences, that are about and from, the essence. Naturalization is about the free expression. This is about the way that God is. He is not for lying, cheating, and stealing, but is about the good. This is about the way the world is about the way the world is about the way the world is for and from, the world. This is about the way the world is about the complex and complete ways that the world is about the way dynamics are about the functioning of the

way that math is about math is about science. This is about also, how language is about writing. These are about the universes, and about the way that it will continue. This is about the ways that the world, will continue, and about how the ways that the world is about the continence, that is about Jesus Christ. He is better. He is much better. And, He is the best, in the world. This is about the way that the world is about the future, and how it is about, the task at hand, and how we must communicate, with the world, that is with the world. This is about the way the world is about the way the world is about the going, and coming, of good, and bad, and about how the world is about the knowledge. This is good and evil. This cannot be compared with each other, yet it is always comparative, to each other, through opposites. These are what do not attract with evil, but do become good, with real good. The "real" good, is about the ways the world are about the way, the world, are about the ways that the world is about the vision, that is about the ways that the vision, is about the complex, and complete, ways that can make a difference, in people's lives. These are about people's lives. These are about Heaven, and earth, and the accumulation of good. This is also about hell, and earth, and the accumulation of bad. This is what was created also. This is after the fall. This is about, the ways that the evil, is about the world, and the good, is about the world. Yet religion could be smart, and sane, and good. This is why I am writing about religion, and its causes. These are causes from the point, of interest, that is about the ways that the point, of interest, is about the accolades, of knowledge, and how they work together. These are to save the way we see and think about things. This is about praise, and communication, and how we must brad, or boast of something, at least, once. These are about the

accolades of knowledge, that we have, and communicate with, and for, and from, and about. These are always about the issues that are always about and for, the issues, that are about, and for, the issue. This is about the issue. This is about the accolades of knowledge. The up and coming, ways that we see, things in the way of, is about the perspective. These are the elements of time. These are the elements of nature, man, God, naturalization, environment, mankind, and the whole of the whole image. This is made in the image, of the context. These are about the context, that are about the challenges, that are about the girl, that are about the boy, that is about the wonder. This is of life. These are about the ways, that the life, is about the life, that is about the life, that is about the life, that is about the death, that is always about celebration. This is because of Heaven, and earth, and how this works. Also, there is a hell, that is also about itself, and the earth, that is about lying, cheating, and stealing, casually, but not intense. The real hell, people go to, for being too bad, through Christ. The heaven is there for people, who are too good, through Christ, and it is not a battle, but a promotion. This is for heaven, that is also on earth, that is also, in your mind, that is also in your body, that is in your brain, that is in other realities, that must be explained through philosophy, yet my physics equation, of this, is about the body of Christ, that is about all. This is about the clever approach, to politics, that is about everyone, and the clever approach to Heaven, is about the creation, of the world, that is through Jesus Christ. This is about the ways that the world is about, the control. This is about, the explanation, that is about the truth. This is also about the trust, that is about also, the factual existences, that are of also about the true and truth in the elements of truth, that are about truth, that are about the truths, that

are about the Bible. These are truths that are about the clever, ideas, that are about the truth, that is about the truth, that is about the ways the truth, is about the ways of truth, that is and are of, truth, and what the truth is. This is about Jesus Christ. You, can think in any terms that you would like, to, be about, and for. These are about the truth that are about the truth. This is about Jesus Christ. This is how we can think, in any terms that are not ours, or ours. There are only two thinking modes, God and man. These are about man, and about how man, and God, are about the conflict, and interest, of mankind and man, and about the ways that man, and mankind, are about the ways that mankind, and man, are about the creation of man, through the cross of Jesus Christ. He died for r our sin nature, to be His. Accordingly, there is a stronghold, in Christ, to believe in whatever you want to believe, in, which is about the real Christians, that are about the way knowing, is about the knowledge, that is about the ways that are about the way the real, that is about the complete, and total, ways that man can make sense, to himself only, which is how naturalization explains all. This is about the ways, the world is about the client, and student, relationship, that is about the client, and student relationship, that is about the ways that are about the conflict, and consistent, ways of thought and apparent, justice, for all. These are about justice, that are about the ways in which we evolve, and see what evolution is afterwards. These are about the styles and generation of thought that is about the conclusion, that is about the remorse, that is about the apology, that is about the sorrow, that is about the gloom, that is about the sadness, that is about the real loss, that is about the real apathy, that is about the real sympathy, that is about the real empathy. This is about the real Jesus Christ, and how He loves you. This is how

He feels. Maybe? Yet, He would give up his life for you. This is with pure and utter joy. These are about the ways, that no one can do things with, that is about the ways that no one can see Christ with, that is about the ways that Christ, is always God, and we are not, unless we see Him, this way, through our own situation. Somewhere, in the world, a child is dying, without Christ, yet this is impossible. All people, will surely know of Christ, in the world, no matter what. This is through the Trinity. This is my goal, in life, which is to spread the gospel, that is through the eyes of the perspective, This is through the naturalization, of the world, through Christ, also. This is about the way the world is about the ways the world is about the humanity, of the world that is about the way the world, sees Christ, that is about Christ, that is about the way Christ, is about the way Christ, is about the forgiveness, of people, or He might feel this way. This is toward the world. This is about the allegory of the patient and the doctor, that makes sense of this. This is that the world is about the way that we communicate, with God, that is through the communication with the demi-God. This is about hypocrisy. This is what this is. This is a person who thinks they are God, relatively speaking. They think they are about the creation of the world, through themselves. This is about good, and about how evil, can become good. This is through the eyes of the hypocrite. This is about how the hypocrite, is about and for, the essence and substances, of man, and mankind, and environment, and nature, and naturalization, and God, and the processes of thought that are in the universe. These hypocrites do not believe in this because they do not believe, in naturalization, but in evil, as a form of "good", that they create and define, and conclude that they are more good, than this, even though they are evil. These ways hypocrites are

about God are about the way God, is in control. Yet, they will not ever realize the truth, of man, and mankind. This is because they are lunatics. These are people who believe, in Satan. This is after all the points I have given, in this book. They do this on purpose, to be evil. This is as good, by thinking Satan, is the winner. They are about the control, and option, of goodness, when they are realistically slaves to Satan. This is due to themselves, only. These are about the ways, that man, is about mankind, that is about mankind, that are about man. These are about mankind. These are also about man. I feel sorry for these kind of people, because they will not ever find the truth. This is to anything. These are the factors of a good relationship, with God. These are how they form. They go through belief in things, that are about you, and not about Jesus, until you discover this. This is to believe. The man in civilization, is about the ways that civilization is about the ways that are about the ways that are about the ways that are about the sane, This is about the way we are smart. These are about the sanity, that is in us. These are about the colossal truth, that is in the control, of the world. This is through Jesus Christ, the one and only God. He can accept anyone. Yet, He does not sometimes, because of evil. This is what some people choose, over sanity. This is the problem of society. This is the conclusion to the doctor and patient allegory. This is how the world is about the way it is for the reaction. This is of the conclusion of the doctor and patient a allegory. The patient and the doctor, is about each other. Yet, they are about the context, that they are inside of, which is about pen and paper. The doctor writes what the patient tells. This is about exaggeration, and completely erroneous information, about nothing but what 1, Satan makes you think, 2. Jesus makes you think, 3. Or what your "freewill"

thinks. These are about the bad, the good, and the ugly. This is how doctors receive the information, that Is through the patient. This is also this way. If the doctor was fair, it was you and him with Jesus. If the doctor lied, it is you and him with Satan. And, if the doctor releases information that is not true about you, then it is with "freewill". These do make sense. These issues of the doctor, and patient, are about those whose issues are about and from the hell, the heaven, or the opinion that creates this. This is unless your doctor believes in predestination. This is always as a free choice. The doctor, who knows, us knows us the most well, in the context of these three. Yet, if you think you are these three, you are a lunatic. This is about the test of doctors and how they see things, through perspective. Naturalization, is about the way, that naturalization, is about the doctor, and patient. These is about the trust, the truth, and the abilities, to comprehend, and also seeing what we have, which is what we go for, in the long run, after seeing. This is about how everyone is like a doctor and a patient. This is about absolute truth. The doctor, in return, thinks of this in the same way. To decide on whether or not the person is sane, he must see into this "future" of God and Jesus and Satan, and the tricks of the world. The real knowledge is about, the trustworthy, and true, ways of saving the world, through yourself, also. This is about the way knowledge and knowing, are about the ways that knowledge, and knowing, is about the way knowledge, and knowing, is about the ways, God is in control, besides Satan who is not. This is unless you believe in Satan. There is also a freewill trick that is in philosophy. These are the things, that people fall for. These are just examples, of doctors. The way doctors and patients are about the conflict of doctors and patients are about the choice, that we have with Christ, or a

angel who fell from heaven, where Christ still is. The naturalization of this would be, man, mankind, and God. This would be the conclusion of God, mankind, and man. It would go like the normal equation. This is

1. God and man is about God and man that is about God. This is about man. This is about God.
2. Naturalization + nature + God + man = mankind.
3. Naturalization and man are about God and man which make up naturalization.
4. God + man = nature.
5. In closing it is God and mankind and man that make up this naturalization equation.

The essence and substance of this, is about God and man and about how God and man are in this is through God and nature. These are the equivalent of man. This is because this is an absolute.

The nature of God and man would form mankind. This would be in mankind's nature. This is consistent of God and man that becomes mankind. This is through the nature of the doctor and patient, and how the essence, and substances, are about the essences and substances that are about the essences and substances. These are about mankind. He is God, man, and nature, that make him up. This is in this contextual outline, that is of the outcome of man and mankind and God, that is about his nature. This would be the nature of God and man, that is through the nature of all. This is because this is absolute.

The God and man and nature of God and man form mankind. This is after following the made-up equation of naturalization. This forms thought and thinking of the essences and substances of life and about living that is about life and living that is about life and living. These are the

forms of naturalization. These are the elements, that form the thought behind it. This is about the way that the form of naturalization is about the ways that the forms of God and Jesus Christ are about the context, that the content, and conceptualization are in. These are in the naturalization. This is about a theory, of physics. This is a theory that makes sense as a language also. This is about how the language is about the way that the nature of God is about the nature of man, that is about the naturalization, that are about man, and mankind, and natures, of man, and mankind, and God. This is about the whole control, of God, and the real Jesus Christ, that is about the Christ that is about the living word, that is about politics that is about the well rounded philosophy of the nature of man. This is naturalization. This is about the way we see God. He is seen through Jesus Christ, that is through God, that is through the context, of the content, of the belief, that is of the communication. This is all about explanation, and the connection that we have in God, to the world. This is about the ways that naturalization, is about naturalization, that is about naturalization. This is always about naturalization. These are the naturalizations, that are about the fact, that are about the opinion, that are about the essences, that are about the substances, that are about the content, and context, of philosophy, and about how the philosophy is about the truth, that is about what we learn in school, and about the government. These are about the languages that are about the ways that languages are about the principles that are about the science that are about the creation, that are about an equation. This is the equation that explains the explanation of God. This is through physics, and the approach of language, to define Him, in human terms. These are about the whole of philosophy, and about the ways that the world is about,

the context, that is about the content, that is about the philosophy, that is about the world. This is about the way the world is about the concentration that is about the philosophy that is about the knowledge, that is about the knowing, that is about the know, that is about the languages, that are about the science, that is about the religion, that is about the physics. These is about the contemplation, of a new art and science form. This is called deformation. This is about the acts, of God, and about how He sees and views, things, that are of nature and naturalization. This is through nature, and man, and God, (Himself), and mankind, and environment, and naturalization, (itself). These are about the elements of our perception, that is about it all. This is about our whole perspective, that does this. This is about the justice, and formation, and the worthwhile, aspects of man, that are about mankind, through God. This is all through God alone. This is about the way the God of the universe, that is about the science, and math, and art, and languages that are in and of, deformation, which is the way you explain the theories of the world, including naturalization, with itself. These are the views, that the new world, that are about, the science and math. These are the ways you can explain naturalization.

One example is:

Man. Man. Man. Man. Man.

Another example is:

Man and God make up the way I see church.

And yet another would be,

God is God.

The real equation explains the explanation in real math and physics terms, that then describes how naturalization works. This is about how the explanation of naturalization works, and how we must know it to be true, which is through deformation, a written language that is possible with anything, including other physics and math equations, This is also a "part of the world" theory. This means that the problem that is about the context, of the world, is about the concerning naturalization and how we know this to be true. This is about the new way that the world, is about itself, being created in a good way, for knowledge, that really has been true always, and in every context, including the Bible. This is about the way that naturalization is about the progress that is about the nature of man. This is about how he interacts with the world, and how he sees the grand illusion, that is part his and part God's. This is about the way, that naturalization, is about the world. It is a creation of the world theory, that makes sense, to the whole of the church. This is because it is a explanation of God. This is about how he works according to theory, This is about the nature of man, and how he sees God. This is through the aspects, of this theory, that we all believe in, because of creation, that is always about the dire, need, to complete and complicate things. These are about the details, of the essence and substance, that is about the reality, that is about the disorder, that is about the way that deformation, is the taking back of the disorder, that is disorder, and making it whole again, through the order, that it was not. This is a creation of the world theory, that is about making new things, come true, through order. What was once disorder, is now organized thinking. This makes sense. We must find

the mistake before we correct it, in whole, in terms of God, and Jesus Christ, and the body. This is of Christ. Christ created the world this same way. This is through the making of the world, after disorder, which means dying on the cross, and coming back as life. The Word of God, was in the beginning, because of this aspect of God, that is "rebirth". The whole world of Him, was reborn after He died. This is about the way the Christ of the world is about the deformation, of religion, and how it teaches disorder, then changes it to order. The definition of "deformation" to be related to my definition, and form, is a formation that is changed, through the stress or strain to an object, as in a movement from one shape to another. This would also mean, a direction that is thwarted due to absolute pressure to the opposite side. The body and direction of the world, is through the language of deformation, through the technical side. This is about the size and shape of the object, that changes, as well as the chemistry, to a new form. This is the form of deformation, that would be the change in the form of an idea. This would be to perfection after learning the imperfect way, that concludes as a "change of shape or form". This change would be directly related to form, and cause, and shape, that would explain to anyone including itself, as a body. This is how the universe, acts, that is about healing, and breaking down, in order to heal, that is naturally deformation's. This is solely about the doctor and patient allegory, that is about 1. Healing, 2. Breaking, and 3. Changing into healthy or related thinking. These are all to others, as in other people, that are the same way as I have explained. To be accurate, the equation, that is about deformation, is about the creation of itself, through the theory, of naturalization, that is about the nature and man, and about its animalistic qualities, that are about his appetite. This is only about

survival of the fittest, again. This could not come from disorder, because of the ways that disorder, is about the discovery, of the fact, that is about the fact, that is about the fact, that is about the fact, that is about the fact, that is about the fact, that is about the fact, that is about the sociality, and traditions, and the honor, that is about the creation of the world. This is how we learn through naturalization. This is about, and for, and through, the essence, and substances, of mankind, and man, and God, that is about the false theory of evolution, but in actual terms, the opposite. This would be the creation of the world, instead of the big bang, and the evolution of man's perspective, which is what Darwin saw, instead of the inclination of being right, that is for, and from, the essences, and substances. These are that which make up our world. These are the fabrics, of society, and the world, that make up tradition, in the church, and through instincts of survival, through the social, and the breakout of the honor, that we encounter because of this. This is about the way that the health of the world is about the way that man and mankind, can cooperate, and get along, and substation the wealth and economic ventures of the entire world, with its essences and substances, that are about the colossal ways, of fighting and living, that are about the substances, of man, and how he can build a world stage. This is about the way the colossal, mankind, and man, builds the tower of Babel, and then falls, until the death of him, or the way that the world is about the ways that the entire world is about Babel. That is because it is only one story, in the Bible, that is also, a allegory, for us to think with, and also absolute truth. This is from the height, of man, that can "take over" and control, the business, and about how the business, is about control, and protection, that is about the fall of man, and how we have that much knowledge.

This is about the way, the world is about the way the God, of the world, of the heaven that is about the world, that is about the decide, and that is about the revelations, that are about the real world, and its substances. These are about the real world, and the way, that the real world, is about the ownership, of the world, that is about the God, about and of, the world, that is about the way that the people, and deciding people, act and will act, about the ways, they are created to act like, and of. These are about the way the world is about the way the world is about the way, the world is about the way the world is about, the way the world is about the way the world is about the citizenship that is in the whole, world, of fantastic, challenges, and obvious decisions, that are about the one thing, that got us here. This is Jesus Christ. This are about the close encounters, that we have with the third kind, that is about alienation, from the sense of the world, that some people, are from. These are fact, and lies, that does not exist, in the world. These could have been created from superstitious beliefs and then the hoax of the entire nation, to believe in the world, as the name "alien" even. This is the term if you are not naturalized, This is the popular belief, that anyone who is an alien, are not naturalized, but have roots, that are from no naturalization, or natural thinking. Yet, even aliens have naturalization, that they must use. This is only if they exist, and yet they do not exist. The nation, that we are a part of, if aliens did have naturalization, must have been made residents, that reside in the country. This is not a copy, of naturalization, that is of our country, but a challenge, that is of natural disorder, that becomes order. This is about the world, that is about the better, of man. This is about superstition, that could have created the alien. There are many people seeing this, and if God wanted to do this because of other

people, then He might have. This is due to the fact, that whatever we are made of, is whatever we believe, in. These are from, rules and regulations, that are about the way, that aliens do not exist, but might be believed in the real way. This is about the way the world is about the way that God is about man, that is about God, that is about man, that is about God. This is about God and man. These are about God and man, that are about the generated, outcome, of the outcome, of God and man, and about how the title of alien, does wrong things to people's perception. This is the superstitious belief. This is that aliens exist. If we are wrong, then why did we believe in the first place? Aliens are strange. Therefore, we have strange beliefs about them. This is about the way the wrong, implication, about aliens, arrives, from lies and stealing. This is about also, cheating. This is about the way the world is about the God, of the world and about how the God of the world, is about the way the God of the world is about the lasting relationship of Jesus Christ, and about how the relationship, is about the control, of the everything, that is about and for, the tragedy, that is of the control, that is about and of, the elements, that are about the control, that are about the substance, that are about the essences, that are about the ways, that of and about make sense. These are part of the English language, that is a part of the literature, that is a part, of the world, that is about the way the world, is about the control, through authority, and how the authority, is about the language, that is about the creation, that is about the created outcome, that is about the language, that is a part, of the universe, that is about the ways, the world has, entitled the issues, of the world, that is about the business, that is about the creation, of the business world, that is about the issues, that we have

prevalent, to the Tom and Maggie, names. These are names that mean nothing.

NAMES AND THE DOCTOR AND PATIENT ALLEGORY-

The fact that is about the client that is about the relationship, that is about the clientele, that are about the relationship, that is about the control, that is about the control, that is about the names, of God, that are also, about man. These are similar, but not in the context of life. The names of people, are made up. This is about how the whole world is made up, to be made after God. So, God given names, are what these are. Yet, parents pick them. This is about how doctors, know people, as their names. These are about how, the people, judge each other, and pick favorite people, to choose from. Then, they make up nicknames if you do not like someone. There is also a concern, about names, and numerology, that is Satanic. This is about the idol worship, and how there is a number. The number of the beast, is the mark of the triple six. This is about the way, the number system works. This is not with names either. One would have to change their name, for one to succeed, for some reason as a criminal, even though all they do is commit crimes, that should not have been around anyway. This is about the ways that the beast, and the mark of the triple six, is about the false prophet. This is about how they changed their thing, to a number, that controlled them. This is about Revelations, and false prophesy, that is about believing, in the beast, that is about, the control, and about how, the control of the beast, is about the client hat is about the relationship, that is about the ways that the challenge, that is about the

sexual orientation, of Revelations, and how angels do not have that,. "When the man comes around, song, by Johnny Cash, is also about the Revelations, in the Bible, that is about the context, and content, that is about the way the world is about the client, that is about the clientele, that is about the kind of people, who would buy this music. Also, my point, is that, the number system could be worshipped, because the one who knows this, is confidential. This is the government. The beast, with the mark of the beast, is about the client and student, of the desire of man, and his joy. This is a prophesy in Revelations. Maybe it has come. And, maybe it will come. This is only about time. The ways that the number system, are about the client and the student, is about the client, and about the way, the client is about the client, that is about the success, that is about the failure. This is about the ways, that man, and mankind, results in anger. Anger, is the last something in which existence is about and containing the formation of this thought, that no one can prevail over. This is but God. This is about God, and how God interacts with the world, that is about the world, that is about the world. This is about the world. This is about the way the world is about the way the world is about the way the world is about the way the entire, world, is about the pure and simple, ways, that the world is about the God, that is about the way the world, is about the way the world, that is about the way the world will be about slavery, if this continues, and will be about stereotypes, if this keeps going, and will be about hate, if this keeps on going, and will be continually about love, if this happens, with the real Jesus Christ. This is about God, that is about the essence, and substance, that is about reality, that is about the ways that we learn, and control the evolutionary theory, and how we deal with this is insane, as mankind, and man, and

God, and environment, and nature, and naturalization, which is why we believe in it. This is because Darwin, was a very strange man, who practiced falsehood, to create this immortal theory. The ways that man is about, the ways that man, is about the ways that man, is about mankind, is about the theory of knowledge, that anyone can practice. This is as long as it doesn't have falsehood in it, and does not lie, or else people would be led astray, almost. This is up to the faith of man. This is able and about the theory, that is about the continuance, of good and evil, that are about the sinning that is about the ways that God, is about the essence, and substance. This is about the ways, that material images, are about the ways that we are about memory, and gaining completeness that seems to be about this economic venture called, salvation. This is in the modern world. This is usually paid for. This is why I am saved. The reason why anyone is saved too, is also dealt with by having money. "The love of wisdom, is the root to all evil" is similar to the "love of money", but only if you think this is about continuance, of the world, and the places, where we can go to, and go to, which competes with the essences, of naturalization. This is about survival of the fittest. This is about, the essence, and that is about the essences, that is about the control, of opinion, and fact, that is about control, and opinionated people, and the factual existences, that are about control and more or less all the other ways, that we can see and call someone "complete". Yet, this is completely dumb, to use a simple expression. This is "dumb" in the eyes of Jesus. He is the one and only, member, of civilization, that is about the ways that the dumb, is about followers, when they do not know how to follow. This is about the faith, when we do not have faith. This is about the outcome, of the written agenda, known as Jesus Christ's plan. This is

about the outcome, of the lecture, that is about the words, that is about the agendas, that is about the ways in which we communicate, that has no bearing, on the real world, but the world where we can understand things. The fact, that the existence, of man, is about mankind, is about the anger, and frustration. This is of the outcome, that will never be realized, that is about the agenda. This is about God's real plan, that is realized only through Christians, that help create other lives, as good and right, with knowledge, knowing their roles, and how they make up, the coordinates, of what Christ wants, and what Satan is all from. These are the ventures that are about the symbolism, that is all about, the repertoire, that is about making babies. This is also about work. This is also about death, that from came the Garden of Eden, This is with the outlook of Jesus Christ. He is the projected one, of the dictation of Jesus Christ. He is the one and only, one who knows the Bible this well. This is because the Bible, is about the conflict, that is about the content, that is about the conflict, that is about the research, that is about the content, that is about the control, that is about the complication, that is about the contemplation, that is about the earth, and its elements. Physics, controls this kind of thought. This is about the fact. This is that there is not a bad or worse God. Yet, there is no kind of hope, without a God, which is why we believe in Jesus. This is about the way that man and mankind and God and naturalization and nature, and environment, and the world are about the created elements, that we do not talk about in the Bible. These are 1.earth, 2. Wind, 3. Water, 4. Fire, 5. Sky, and 6. Air. These are the basis, of a theory that contains these elements. These are:

Environment + Naturalization + Nature + Man = God

This is the explanation of the six elements that could make up our universe. This is about the way that the world is about the elements that are about the make up the earth, through science. This is about, the essence, and substances, behind the elements, that are of the atmosphere, and creation, and building of the universe, through these, which is, in theory, the six things that made up the structure, before the big bang. Yet, the other way of seeing it, besides the "big bang" is the Biblical interpretation. The only explanation, behind the two is that the elements, that created the world, are false. They come from themselves only, because of the way the entire chemistry of creation, is through elements? This is the good objects of and, that occupies, all time, and space, and is about the big bang. This is about the big bang theory, in which God, which is of that is always about the chemistries, that is also about the way lying is the opposite of creation, with God, in whole.

One person could "lie" and say that the world, was created, through the elements, but it is really God, who created them. This is the only fallacy, of the big bang. This is that Christ created the world. The only scientific explanation is about the chemicals that is real, is that the chemicals must have created themselves, through the occupying level of advancement that consists of and concludes of a force that is stronger than God. He is the one who is in control, of the world, and this must mean, that the elements are created from a bang. This would be an internal, external, mishap, of these elements, if this is true. The elements of fire, water, wind, earth, air, and sky are somewhat connected, to the beginning, in my opinion, before God. These are the opinions, of man and mankind that do make up naturalization. This is a theory of God,

which is about man. He is not the one who created the world. Man, just sometimes is wrong, about God. God is the sole creator, of the world, with the world, that is about its own creator. He is God, and is living, in the God, way of the world. This is through, how explaining the world is about the creation that is impossible to explain in science. This is unless the elements are about the makeup of the creation of the world, through God's voice, which is pure hypocrisy. This is the language of the creation of the world. The elements are about, the construct, of the world, naturally, that is and are about the creation of the world, and the makeup of the society that is about Jesus Christ. This is about the church body that is about explanation. These are the church, body that is about the makeup of everything church. All the rest is the government. This is about the way, the world, is about and through, the client, that is about and through, the social issues, that are about and through the science, that is about and through, the dynamics. These are of God, or man. These are the explain and know, theories of man, and how he becomes, with his life, which is about the ways that man, and mankind, are all about God, in which God, is explained to him, with factual existences, for some reason. This is the reason that we all know. This is of faith, through Jesus Christ. But, what are the elements, of faith? These are the functioning, and functionality, of the Roman Empire, before it fell. This is all about intellect. Yet, the modern day America, cannot fall, because it is made of the fabric of Jesus Christ, which is almost known as the real "Jesus Christ" known as the real God. This is also His Father's position. This is what comes to the doctor, patient, and allegory. This is of and about, the ways, that mankinds, and man, are about God, and nature, and naturalization, that is about the environment, that is

about the good and bad, options that we have as a people, of a Christian faith. This is about and beyond the knowledge that is of and about the relative positions of power, in the world. These are made up positions that are about the essence and the substances, in which are about the philosophy and theology, of the Christian faith. These are the direct, options that are about writing what we want. Everything we write is our own opinion, based on what we learn, and control, and know. The option of knowing, everything, is about the faith, we have in Jesus Christ. He is the chosen one. This means that the world, and God, chose Him, known as the creator, of the world. This is through His Father. He created the world also, because of Him. The real and unknown aspect, of God, is about the knowing, and showing, of God. This is about the way the world and God are about the ways that nature and naturalization, is and are about the client, and his special relationship, with his boss. This is what the doctor and patient allegory is. It is God with us as his followers. This is a meaning, meaning, that God can forgive man, and give him opportunities, to also make and believe, the Word of God. These are through our actions, and motivation capabilities, that are about the real God and man, and what is about the way man, and mankind, are about the control, and manipulation, that is about the freedom. This is what freedom symbolizes. This is man. The real one true, way, of God, symbolizes all. And, theory symbolizes theoretical thinking. This is and always is about the construction that is about the made up society that is about the ways that mankind, is about man. This is about the way man, is about the strong, and about survival of the fittest. This is about and because of Jesus Christ. This is about Jesus Christ, that is always about the contents and context that I have been talking about. This is about the

religion. This Jesus Christ, figure, symbolizes all things that are all things that are all things that are about and of God. He is of the Son that is of man that is of mankind that is about and of, the nature, of man, and mankind. These mankind things, which make up mankind, are only about mankind. These man, things that make up man, are only, about man. These are the God thing of faith that makes up God, are only about God. These are also all the things that make up each other, in sane, relationships, with others, in a world. These are about the people we meet, and the way that we are about this. These are about the ways that are about the ways that are about the ways that are about the ways. These are the constructions of justice. These are the reality, and all inclosing examples of all. This is about evolution, that is about and are, about and are about and for, the example, of all. This is Jesus Christ. These are about the ways that the control of Jesus, is about the control of the world. He is who is about the way of that we think, and feel. He is the one who is the Word of God. This is about the ways that the helping, of man, is about the concentration, of good and evil, that is about the knowledge, in is of which that is of the helping of good. This is against the opposite of evil. This is only good. The fact, that I like the ways, that help is about Jesus, is the fact, that Jesus Christ is alive. This means that there is an eternity by the way He created the world. These are about and of and are, from and with, the creation. The creation is God's, Jesus Christ's, the Holy Spirit's, and the Voice, and also the Scripture's. These are all contents that lead us to religion. This is my conclusion. This is about the ways that the conclusion is about the always prevalent ways of the magnet. This is all. He must have had a strong magnet, placed to Him, in order to create all of this. This is through the utmost. These are

about attitudes that are about the conclusion, that is from and about, the symbolic meaning, that is and are, of Christ, and how He communicates. √ This is about the ways that man, and mankind, is about the control, and construct, of the daily planner, known as man. This is also triumphed by God. This is if we believe in Him. √ He is about the God, that is about the God, that is about the God, that is about the God, that is about the control, of the God, that is about the relationship, with the God, that is about the control, which is of God, that is about the control, that is of Jesus Christ, which is God's. These are all about the control, of the substance and essence, that is of and about, the know, that is of and about the truth, that is of and are, about the ways that the language, that is about the ways that the world, is about the ways, that the elements are about the configuration, of the world, and about the ways that we control, the world, is through the direction the world is about the context. The minerals are about the contextualizations that are about the realization, of the logic, of ours, that means that it could be the creation, of the world, through logical treatises. No man can understand what created the world by the way Jesus Christ, already and is only explaining this. This is always a debate with scientists though. √ The theses so called, exist, become and are scientists, that is about scientists that are about scientists. Most are Christians, which is what, the world does. If so, it is because of Jesus. √ The holy aspect, of God, is through the ways that the world, is about the way we are, about and of, and are become, with and about. These are the small fractions of percentages. All of these people, are Christians, with the same logic, as I have demonstrated, in this book. √ The book, is about the way, the world, is about the way, the world, is about the ways that the world, is trustworthy and true, for Jesus Christ.

This is about the ways, that mankind and man, are about the triumph, in which is that is about the clear, and utter danger, of not being a Christian. Christ has planned out the way of the Christian, and others do not know this at all. √ These are about the ways that are about the conflict, and proposal, that is of and about the ways that we can connect, to the outer world, through Jesus Christ. √ This is about the way, the world, is about the way the conclusion that is of and about the ways the man and mankind, do well. These are always with mankind. √ The content of the world, is about the ways that the world is about the ways that the world is about the way that the world is about the way the world is about the way that the world is about the element, and naturalization, and God, which make up the whole world. These are of ancient Chinese philosophy. My point, is, that no one can create the world, but God. Yet, these are close reasons, to see why God, does things. He could have been created, by thought, from Jesus Christ, and then made it the beginning, of the real world's creation. This is why He might say that the world was created by Jesus Christ. This is also, through the Jesus Christ Trinity, the one and only true God or God the Father. These are the creators of the universe, and Jesus. This is just a matter of time. √ The way we solve insanity, is by exploring into the realms, of others. The people, who are the "creators" of their own "lives" are really without God. This is my one true point to exist. √ In existence, there is a true creator. He is God. √ This is about, how God has a maker and His name is Jesus Christ, and how Jesus Christ, has a maker, and he is His Father. Yet, Jesus Christ saves. √ And, God the Father, reacts differently, to the own ways of living, life to the fullest. √ This is about the ways that God, and man, triumphs, over evil, and Satan. This is through hell. √ The way that the fact is about

hell, is about the way the world is about heaven, and the opposites that do attract. These are e the reasons that are the way the world feels about this. This is because Jesus Christ works like this. This is about the way that Jesus Christ works, that is through the language of thought, and thinking, that is through Jesus Christ Himself. This is through all thought. . √ This is about the way the world is about the obvious, and about how the obvious, is about the condition that the world is in. This is about, the condition, that is about the way that the world, is about the makeup of the world and about how and when the world will make it. This is through the world, view of the world, and its violence, and hate, and anger, and about how this, like the ways also described in this book about God, and elements, are about the living truth, that is in the Bible. This is about what the Bible, says, that is about what Darwin says, that is about what a judge in heaven known as Jesus says. This is about the violence, and hate, and anger that is about the ways that man is about God. This reason that is about the ways that we are about the way, is always about Jesus Christ. This is only. . √ The reason why, the world, is about the progress, that is about the world is about the directed outcome, of the progress that is about the ways, that are about the ways, that are about the philosophy, that is about science, that is about physics, that is about the entire world, is about the made up philosophies, that are about the progress, that are about the direction, that are about the ways, in which we see things, that is about the ways that we learn, that is about the direction that we follow, that is about Heaven, We follow what we do not know. . √ These are about, the ways that are about the concern and justification, of the follow of the leader, of the doctor and the patient, allegory, of God and His Son. His Son, is about Jesus Christ, which is about Jesus Christ,

which is about the concern, for the love of God, which is about concern that is about McDonalds. These are about our favorites. This is the real theory of God and how He does not pick favorites. He saves. . √ These are about the ways that natural, things, are about natural things that are about God, and Jesus Christ. √ These are about, the fact, that is about the fact that is about the factual existences. These are fully about Christ. He is the creator, of existences, that are about the philosophical, treatises, that is about, the concern, that is about the ways, that are about the philosophical, that are about the recollect, and memory that is about the science, and math, of man, and about how mankind, has been included in a science and a math, always. This is also how the world is about the ways, that the world, are about the ways, that the life, of the world, is about athletes, and the way the world, is about the way, the world is about the way, that the life, within the world, is about naturalization. This is about not naturalization, but the evolution. This is of nothing. There is a big picture, that is about the way, that the bigger picture, is about the philosophy, that is about the client, is about the relationship, with the client, and student that is about the results, that is about philosophy, that is about the way the philosophy, that is about the ways of the future, that are about the collective, and generalization, of the world, that is of and about the ways, that actors and people, see each other. This is totally different. This is, as a whole, how the world was created. √ This is what the world is about that is about the way the world is about, the ways the world is about the cross, and the forgiveness, that is about the conclusion of God. He is the one who also saves. This is by, the way, Jesus forgives a person that goes to God. This is about how God is about the truth that is about the truth that is about the truth.

This is about the aspects of truth. √ This is about the way the world is about the heralding of the world, to the naturalization, of man, and God and how we think. √ This is about the way, the naturalization, is about the ways, that the conclusion, is about naturalization, that is about the client, and student, relationship that is about the ways the relationship is about the ways that God moves in us. This is for the government, school, and church. √ This is about the way the world is about the way the world is about the way the world is about the way the world consists of nature and man and so on and so forth. This is the contents of the complexities of naturalization. This is about the whole existence of God, and about how Jesus Christ forgives police officers, and every most powerful person. This is if they are not corrupt. These are the stereotypes of the most powerful people, as well as police officers. This is because people stereotype them as corrupt. They are not corrupt though. This is due to God. They can be forgiven by God, known as Jesus Christ, also. These are about the ways that, Jesus Christ, is about the God, of the world, that is about the ways that Dallas, Texas that Is about, the killing of innocent criminals, and how the justice system is corrupted by Senators, and Congressmen, and the President, if we choose to believe in this, this way. The congressmen, of the United States, are about the Senators that are about the Congressmen that are about the powerful families and how they are a part, of these. These are issues that are about, and of, the issues of policies, of the family, and how they react. These are to the issues, of family, and friends, and explorers that are eventually of the Christian faith. The church is about the corruption that is about nothing, but powerful people. This is always where it stops. The church of the world, makes up Christ's body, according to the Biblical interpretation of

mine. This is always true. It doesn't really matter, what you think sometimes. Yet, interpretation is about the way the world is about the way the world is about the way the corruption, is about the ways that corruption is about the ways that corruption, is about the ways corruption, is about the ways corruption, is about the ways, that corruption, is about the ways that corruption, is about the ways that corruption, is about the ways that corruption, is about the ways that corruption, is about the ways that corruption, is about the ways that corruption, is about the language, that is about the ways that corruption, is about the ways that corruption, is about Jesus Christ. This is about blame, sinning, contemplating wrong, doing evil deeds, deciding on evil for your life, reading the Bible wrong, seeing into the eyes of Satan, doing very wrong things, to everybody, and seeing lies and cheating and stealing and not learning from them, also not reading the Bible, and going to church, as well as reading into the future wrong, as well as being anything that the Bible thinks is wrong. These are all true. This is even if you think that you are not corrupt, or doing these things. The Nobel prize, is about the corruption, of Satan, that is about the corruption of powerful people, that is about the corruption of the nation, that is about the corruption that is about the powerful elite, that is about the corruption of, wicked and sinful people, that is about the dialect, and reason, of the intellectual aspects, of man and woman. Some people do not see corruption this way. These are the Nobel Prize winners. They can resist the world, but also not risk corruption. This is only if they change it their way. This is by even having to be corrupt, sometimes. This is with a guilty conscience, and good attitude, toward the not corrupt practices of the church. These are about rape, if you want to do away with it, violence, if you want

to participate in it, to see how to stop it, and lastly, saying that the wrong of your country, is yours. This is by investigating people, around you, sometimes, and maintaining a leadership role like this, which understands the wicked. These are the roles of civilization that are about the roles, of young life, that are about the Nobel Prize winners. The Nobel Prize, I would like to put this book up to, known as the Nobel Prize, in literature. This is because I am explaining God, known as the real Jesus Christ, and how He works. According, to the Bible, these men are evil. But, I see a brighter day. I see people seeing into the future, and knowing what it will bring. It will bring a better day. √ I have seen, the dawn of the ages, in the written form, of knowing and seeing, into the future, for my own cause. I think people, learn their lesson, by seeing, and actually participating in wrong, at least once, and then the Lord and Jesus Christ, come to judge. √ This is by the real Jesus Christ, who washed away all my sin, for repenting this much. This is why I am a scientist, of the Christian religion. √ These are about the elements that are about the progress that are about the dynamics that are about the beautiful people that are about the colossal. These are about the ways that we are about the digression, of the human faith, and how we are humanity, that must believe in Jesus, in order to survive, and make known, the Nobel Prize. This is also, an institution that excels, and knows about the dialectic, and issues, of the humanity. These are for the world. √ This is about how the world, is about the religion, that is about the religion, that is about the religion, that is about the religion, that is about the religion, that is about the religion, that is about the religion, that is about the religion, that is about the religion, that is about the religion, that is about the points, that are about the humanity, and where this comes from, in essence, and ap-

peal, of justice. This is the appeal, of the justice, of the world, that is about the ways, that are about the ways, that are about the way, that is about the way, that is about the way, that is about the way, that is about the way, that is about the religion, that is about the choice, that is about corruption. You can choose to be corrupt, or not corrupt, by going against the laws of the land. This is about the way laws, are about the language, that is about the sense, that are about the issues, that are about the senses, that are about the essences, that are about how we learn. These are about the prizes that we go for, and try to achieve. These are about the senses that are about the dialect, that are about the language, that are about the senses, that is about the progress, that is about the senses, that are about the language, that are about the titles and positions, of the world, and how these are sometimes, strange. These are about also, the, issues that, are about the issues, that are about the issues, that are about the progress, that are about the senses, that are about and from, the issues, that are from a good and noble heart. These are about living, behind the heart, of Jesus Christ. When you take an oath, you take this to heal, and make good, a life that is about yours, and ours, and others, in the name of God, which is on the Bibles of Jesus Christ. To make, a name for yourself, you must look to God. He is the provider of good and excellent things that are about the senses that are about the commitment, that are about the life. This is what the books of the world do for you. I am writing this in my name, as well as anonymous. $\sqrt{}$ This is about the issue of life that is about the issue of life. $\sqrt{}$ These are the parts, of life, that are synonymous, with the religion, of man, and how the religion, of man, is about the opportunity, and which are about the cause, that are about the senses, that are about how we were created. This is in order, to please

God. √ All the quotes in the world, cannot suffice to this. This is even, if the world is about danger, and escape from this, in whole, and through the complete. √ These are the natures of man, and how they interact, and involve themselves with man, and God. √ This is about God, and what God, is about is about the death, of senses, and how these were gone, of Jesus', when He died. Yet, He rose again. He was not senses. He was God of all. √ The essences, of the world, are through the dynamics, of what we know and what we understand, about life. √ This is about the way that we know about, life, and what life, is about, is about Jesus Christ. This is about Oxford University, and how I have high hopes, in this, essences and substance wise, way. This is about the school, which is good enough, for the real whole entire government, which is about the hiring of people. This is based on how big of a history that we have. This is about, the way that the world that is about the way the world is about the way that the way the world, works, and are about the ways that the world works. This is also, but about, in addition, the substance. This is about, in addition to for, the essence, which is about the essence, only. This is about, the ways, that the whole world, is about the containing, matters, which is about the ways, that is about the ways, that is about the ways, that is about the way that is about the ways, that we have substances, and essences, in us, and about us, and from us, to the real world. These are the code words that are about the ways that is about the friend that you have. Some people think their friends, are their "life". These are about the ways that we are about the synopsis, that is about the ways, that the find, and go seek, are about the "hide and go seek", which does not make sense. This is unless you know this thing, and this fact, of information, that is about the ways, that the open, and the closed, are

about the knowledge, you accept, and the knowledge you do not. This is compared, to everything, as a whole. √ This is about the way, that materials, are about the ways, and means, that we survive, with, that is about every theory, almost. These are the theories that are about the ways that is about the ways that are about the ways that are about the ways that are about substance, and essence. The ways these elements, are about the progress, of man, is about how God controls these, also. This is the outer spectrum, of perspective, and how we deal with knowledge, of our outside world, through elements. We were designed with these, and by these. This makes sense. This is about the ways that the world, is about the ways, that the man and woman were created together, and how they intertwine to know this. This is after the fall of man, but was not. These elements are what God, created the earth with. This is about the way that men, and woman, are about the senses that are about the senses that are about perception, because of these elements. These are the industrial ways that man and woman are designed to think, with, and act with, and live with. These are about the ways, that man and woman, live with, and act. These are about the ways that we act. These are with what we learn, and then how the world is designed by elements, that make sense, to man and woman, which are about, the elements, and how God structures them, through His creation, or I would be lying, about all. √ These are about the ways that man, and woman, are existing, but also is about the essence, and substance, that is about the senses, that are about the elements, that are about the control, that is about the construction, that are about the sense, that are about the sense, that are about the Godliness, and that is about the Godlessness, that is about the elements, of the desire, that is about the con-

tains, that is about the context, that is about the realization, that is about the ways, that man and God, are about God. These are about the ways, that man and woman, are involved with these elements, in them even. These are both extra and introverted, and have control over the earth, the wind, the fire, the water, the sky, and the air, and how they interact with these, could not ever be explained, on the basis level, unless you were God. A hypothesis, is that God formed, from these elements, also. He must have known that the world, is about the ways that is about the ways that is about the way that these form, because of Him. √ No matter, what you interact with, and control, there is always an elemental issue, in the world that affects you, but really does not. These are about the ways that man and woman, are about the ways, that man and woman, occupy time. They also occupy earth. We have knowledge of how things work, through natural knowledge, of these elements of the earth, also, which created them, perhaps. Or, is it that God, created the elements, for us to not know about, and interact with? These are the elements, that are about the earth, that are about the essences, and substances, of man, in this theory.√ This is about the way the world is about the way that the world is about the way that the world is about the ways that the world is about, the ways, that the world, is about the ways, that the explain, and know, is about the knowledge, of what is and what is to come. These are about elements that are about the God that is about creation. He does not explain, in context, how things work. This is what is taken by faith. These are the faiths, of the world, that interact, with the world, and the way, the world, reacts, and essences and substances create and destroy, the life, that we live, with, and from. √ The way, the world is about the construction, that are about the es-

sences, that is about conformity, that is about the context, that is about the contextualization, that is about the knowledge, that is about the knowing, that is about the knowing, that is about the ways, that man, and beast, compete. This is through work. √ Work, is designed, to show us, that we are God's property, like beast is man's property. √ This is about the ways that the world is about the ways that the world is about the ways that the world, is about, the way, that the world is about, the way, the world is about female, and male, and according to the Bible, were created together, √ This is about the way that man is about the way that man is shaped by the elements, with his perspective. This is about the way the world is about the way, the world, is about the way the world is about the perfection, that is about the perfection, that is about the perfection, that is about the perfection, that is about the perfection, that is about the perfection. This is about the elements, and how they do involve themselves into, the world, and our perspective, and the ways that perspective, is about the generalization, of the elements, and Jesus Christ. Why do we eat, the "elements" of the body of Christ, if there were no natural elements that created the world, abroad and in specific, with Jesus Christ. He did not just think of this, as an example, but as a real thing. √ The elements, which are about the body of Christ, could also be the real elements, which were in the beginning, and this is where Christ learned, to take the bread, and blood, and make Himself everyone. This is solely in the church, and where it is in the church, is in the Eucharist. This is about the way, that Satan, could have in the elements before the world was started, that have to do with, things in the modern day, which are about different elements, than before. He could have fell from Heaven, with these elements, that are about the elements, that are

about the elements, that are about the elements, that are about the elements, that are about water, fire, air, earth, wind, and sky, that are concluding, to be a part, of the world, that we are not a part of, but within the knowledge of all, without Christ, Satan could know something like this, in essence, and substance, between the real and the fake, but with himself, only and only because of himself. This is about the way that the world is formed, according to the big bang. This is from an explosion. And, there must be elements, in this explosion, to create the earth. These are fire, water, wind, sky, earth, and air that must be completely true, with Satan only. He is the opposite of the Trinity, that has to do with Christ's body. This is only Jesus Christ? He is the only, thing, that has a body? No, Satan is the owner of great societies, that have always been around. He is the "king" of the world, that is about the elements, and how the world was designed, for him only, through these only because of the way he fell, from Heaven, and came to earth, to know this. √ The elements of the earth, are like him, in that they are naturally there, no matter what, and fall naturally, from Heaven. This is where the big bang theory comes from, which is about the creation, before the world was created. This is from the elements of these six earth elements, and about how they formed, also. This could have been, to create our outer world. No one knows, how powerful Satan, is, and these people do not know why , we see knowledge this way. According, to the Bible, Satan fell because he tried to be smarter than God. These are evidences, that he could have been the creator, of the elements of earth, that do revolve around the sun. These are the elements, of the universe, that are about the creation, of the world, that are about the creation, of the elements, that are the elements, that are of and are about Satan. He is the one, who rules

the world, who deceives everyone, and we already know that he is consistent of knowledge too. How can he rule the world, though, if he is not part of the elements, of the earth, like Jesus is part of the elements of His body? This is through knowledge, that formed, because of these things. These six elements, make up the knowledge, of how Satan fell, and rules the world. These are the element, of the outer world, that makes us all smarter than God, but not realistically. God is the creator of these, and this is the Biblical explanation, to evolution. The world, is smart, also, due to intelligent design. These are about the creations, of thought, and smarts, and the elements that make up our world, are part of the spirit of God, that was "hovering over the waters" and then somehow spoke, and created the world? These are also the elements that must have been created, not just through a "Spirit". These were water, air, wind, sky, earth, and fire, that must have been around, before God. This is true, and obvious, or I would be a liar. These things that are about, God, are about man, and mankind, and about environment, and about naturalization, and nature, and these six elements, that make up the chemistry, of the world. √ These are the elements, that participated in the big bang, but are not about God, and His voice. This is the only way I can explain the false, theory of the big bang, without God. But, without Satan, there might have been a voice, alone. This explains physics. √ This is because the voice, was God, and the voice, was with God, in the beginning, known as the Trinity, with God and Jesus Christ. √ Yet, the Eucharist, explains this better, and this is with truth, that is always lasting. √ These truths, that are about God, are about how God, is about how God, is about how God, is about how God, is about how God, is about Himself. √ He is creator, of the universe, as well as these elements, that were always around.

The spirit spoke, and hovered over the waters, and divided the night and day, and separated the waters, from dry land, and created man and woman. So, the elements, according to the Spirit, are about the ways, that the water, and hovering of the Spirit, was absolute truth, and according to doctors, was totally "sane". This is the doctor and patient allegory that makes sense, to this kind of thing. People believe, in health, because it was a creation, of God, but in reality, the patient did not ever, know about the health, or the doctor, until it was discovered, by both. This is because the doctor, created good health, but the patient is who he created it for. This is about the essence.

The essence, is spirit (fire) + water (energy) + earth (God) + air (intelligence) + wind (obvious) + and sky (truth that has yet to be discovered) which is known as the trinity of Satan. This is before he fell from heaven. √ The way that we know this is true, is through naturalization, or the earth couldn't form without it. This is basic fact. √ These are about the facts, that are about the nature, and man, and naturalization, and mankind, and God, and environment, and the factual. These are with the existence of these. These are the basics of physics and science, or I would be lying. √ These are about the facts, that are about the opinions, that are about the outcome, that are about the reason, and rhyme, that is about our planet earth. These are also of the universe. These are the people, who believe, this, who are about the way, that the planet forms, with the same elements, as before. These are about the triumph, or intellect, and about the wisdom, or water, and about the belief, or God, and about the obvious, that makes up fact and fiction, and the spiritual prowess, that makes up fire. This is also in alchemy, but to the extent that it is wrong. This is to the church. This is what the

earth, is made up to be, without God's naturalization, and just with the naturalization, of the world. √ This is about the ways that the world, are about the God, and Jesus Christ, of the way the world is about the client, and student, that discovers this through the game. This is only through the "game". This is not the truth. The truth, is hidden, and no one sees it, known as sky. This is because the "sky" is the most protected, and trustworthy, of all the elements, that does not do anything wrong, and does not cause destruction, of the precious force, behind the ways, that the world works. These are always through God. √ The ways that the world, is functioning, is through the elements, or we would have died. These must be sacred, and protected, or the world goes away. These elements though, were created after Satan fell. They are 1. Death, 2. Life, 3. Pain, 4. Energy, and 5. Future, and 6. Past. These are how Satan fell from the Heavens, and created all of this by "walking on the earth" as a "death angel" with "free will" and "lies, cheating, and stealing, and "evil" as a real false God. He is the one who made man fall. He is the "serpent" in the Bible, which is about lies. He is the "False God" in the Bible, about stealing. And, he is the "death star" in the other religions, that comprise those religions, into being false, supposedly. He is also, into alchemy, and through sorcerers can create things with magic. Yet, alchemy was always disapproved of by the church, which is about the explanation that God does not exist, so the church is like a real God. It is actually the body of Christ. The "body of Christ' is all these things, that was also in the beginning. √ The Christian naturalization, explains, that we are all part of God. This is Jesus Christ. This then, makes the real Jesus Christ, shape all things, that are in the world. But, this is also through God, in the Old Testament. To get to Him, one used to have to perform rituals.

Why is that? This is because history, is a total lie. √ The real reason why we do things, is through the Trinity. Or, we would have all been apes. √ These theories are about the creation of the world, that is about the way the world that is about the way the world is about the way the entire world is about peace. This means that the world is created by Jesus. There is no such thing as a "false God" though, or is there? This could be "Satan" or the "personification of evil", and how he rules the world is through the lies. He could lie about the world, but he would be there with it. People like Darwin do not give into the theories of the world. They just write them. The elements of the world are created, by Jesus Christ also. Then, there must be a false God, directing us to war, and evil, and violence. These are the containments that really create the order, of the world. These are the options that are about our advancement, that are about the creation, of the world, through the client, and the student. This means that the doctor and student allegory is true. The doctor does not create he world, but can participate in it, due to his patient. Yet, what the patient believes, is what the doctor believes. This is that lies, and cheating, and stealing, are true, according to Satan. He is the one, who creates the world, as his. He is the one who creates the periodic chart of elements, yet in a strange way. This is through his "false God" power. This is through deceiving the real God, through lies and manipulation. These are about the ways, that science, and math, and physics, and English, are about the ways that Christ is truth. These are the ways that life, is about the ways that we have reason, that we have opportunity, that we have always had time, to have time, to have time, because someone had to discover it. This is how I discovered naturalization. This is through the ways, that man, and mankind, and God, and nature, and natural-

ization, and environment, have gained a place, in the world, through my writing. This is all completely true. √ These is about the way the world, is about the God, of the world and how He communicates, and reacts, with the way that man, and Him, create ourselves. We were created in God's image. This is of "Us". √ The fact in which is about evolution, is about the ways that evolution, is about the recollection, of a theory that does not make sense. This is about the way that the world is about the way the world is about the justice, of the world is about the way the world is about the God, that is of the world, and is of the world, that is of the world, that is of the world. This is of the world and about the way the world is about the way the world is about the process of knowledge, that is in the fact, that the world is in the fact, that the world is in the fact that the world is in the fact, that the world is in the fact, that the world is in the fact. This is about teachers, and students and how teachers do not really teach, their students, but really teaches, them in whole, the fact, that is of the fact, that is of the fact. This is of evolution, that is of naturalization. This is that teachers are wrong and students are right. This is in whole, the fact, that teachers, are about the client and student relationship, that is about the job, that is about the whole, that is about the opinion, that is about the teacher, and student relationship, and about how we get along. This is about the way, that the teachers, are about their students, that are about teachers, and that are about students. These are students, that are, about students, that are about students. These are about students, that are about the students, that are about the students, that are about the students, that are about the students, that are about the lies, the cheating, and the stealing. This is about social standards and how they, are about, the social standards, that are about the process,

and that is about the standings, that is about the complex, that is about, the client, that is about the relationship, that is about always God. He is the "great provider" of all, the things that are good, and with bad things, He conquers. He is about the way, the world is about the way, the world, is about the way the world is about the function, that are about the way that are about the social standards, that are about church, that are about government, that are about school. These are the judgmental teachers. These are the ones who teach, and excel in teaching. These are the ones who are about teaching that are about teaching that are about teaching. A bad teacher does not deserve to know God, because they choose to not know God, in this very same manner. People who do not know God, do not really know God. This is how the doctor is to the patient. This is only about sick and not about healed. This is through the whole, family and the whole society, that one is a part of. These are about people, who do not decide. They are lost their whole lives. Yet, this is about teaching. They do not teach, with Jesus, but teach on their own. This is for hypocrisy. So, there must be a program that involves hypocrisy, with the world, as an "I'm third" kind of thinking. This is of God first, others second, and self third. This is the opposite of hypocrisy, but if one person decided to be this, with hypocrisy, they could be the real Devil. This is about how good people, know the Jesus Christ, and how bad people, look like they are putting others, before themselves, when they are really condemning people. This is through hypocrisy. And, the love of Jesus knows no bounds. He is about the client, that is about the student, that is about the relationship, that is about buying a Bible. All one has, to do, is buy a Bible, from a store, and gain salvation. This is through the free gift of life, that is solely in Jesus Christ. He is Christ the Lord. This is

about the way, the future, is about the way, the past, is about the way, the future, is about the way, the past, and is about recollection. This is the memory process. This is from the process, of the government, that is about files and records, that do not exist. This is unless one person, does something bad, to another, A crime must happen for someone to be arrested. A person must know things, before he writes a book. A student must be taught, for him to know something. Yet, something is about the student, that is about the teacher, that is about the student, that is about the teacher, and the knowledge that is shared, between the two. These are about the setup and how the client and teacher are about the processes, that are about knowledge, that are about client and teacher, that is about the corruption. This is about power, in school, and how we are taught that power is corrupt. This is because of school. But, school is a new frontier, that cannot be explained, to me, or others, unless it makes sense to all people, in the world, in the world, in the world. These are an explanation, to the process of thought in deformation. This is that 1. In the world is about the whole world, as a separate entity, that is without the whole, world, that is without the whole world, and 2. The process of believing in things, are about the processes about the ways that knowledge, is about the recollection, and 3.the whole world, is about the ways that the world understands things, that is about the whole world, even if it is mentioned three times, and explained three times, until it makes sense. THE art and science of deformation, is about the art and science of deformation, that is about the art and science of deformation. This is about explanation. The first, part of deformation. The second part of deformation. The third part. These are about the world, about understanding, and about art. These are about art, and

the science of explanation, and the understood, and the real Jesus Christ.

You can only understand, things, with Jesus Christ. This is and always is for and about, these ways of exploring, and understanding, the factual existence, from the factual truth, that is in the factual existence. These are in the factual truth. This is about the ways, that the image is about the ways, that the world is about the honorary title that is about the truth, and its explanation. These are about the codes of our universe, and how he and we function. These are through naturalization. These are of and about the ways, that naturalization, is about the finest, and most glorified, and most triumphant, of forms. These are all that can be taught.√

OUTLOOK AND THE FACTS OF NATURALIZATION-

The God of the world, is always about and for, naturalization. He is in control. This is about the way that the smarts of the word are about the conclusion that man draws at the end of his life. This is about whether to decide on God or not. This is actually one that is always around. These kinds of people, are about the ways that God has shaped their life. This is about religion, and how the living times of religion, are about the context, that you choose it in. God created naturalization, and created the world, if you choose to believe this. But, God's free gift, is for anyone, who chooses to comply with it. This is about the ways the life of the living, is always brought to life. √ These are about, the controls, of life. These are about the decisions of life. These are about the conclusions of life.

These are about Jesus. In the end, it doesn't matter who knows about you, in essence, and substance, but who knows. This is always Jesus. He does know. And, He has always known, ever since day one. He created the universe, and He knows about the difficult, way, of life, and living, that is about Young Life, K-Life, and other charities and good organizations, that are about the way we live, that are about the ways, that Christ, in man, is about the decision already, that will last forever. If you choose to accept Jesus Christ, then you can have eternal, life. This is based on a decision. √ This is to save yourself. Or, this is to become a Christian. These do not both work, one must have a savior to live eternally, with God, in Heaven, in the Paradise o f the Garden of Eden. This is about the life, of the environment, that is about the life, of the world, that is about the nature, of God, and man, and the nature of Jesus Christ. √ This is of the whole of Jesus Christ. √ This is to be of a part of Him, known as God, known as Jesus Christ. He is the one, who answers, all your questions, and can approve or disapprove, of your salvation, but usually is a free gift. This is from everything, √ The nature of God, is in the nature, of man, and the nature of man and God, is in the control of salvation, and the processes of thought, that is in thinking, that is about and of, thought, and thinking, and the relationship, that we have in God. This is through the Eucharist. √ This also shows itself, in the form, of baptism. This is also, of the control, of the body, of Christ, and about how the Christ, is about the relationship, between God and man, and about how the God, and man, are about the concern, of the followers. These are with Christ. This is forever. This is about how, the 1. God, of 2. Man, 3. Wasn't actually created by naturalization, but 4. Explains it very well to man, known as the 5. Theory of Naturalization. This is about how explanation, of the body of Jesus,

is about love, and works, and goodness, that all comes from faith. √ This is about how the love, and goodness, and triumph, comes from 1. Winning the race, 2. Keeping the faith, and 3. Accepting God's plan. This is about the fighting of the good fight, also. These are verses, that were created from Jesus, and His body, that is about followers, that are around always. These were also "before time" if you choose to think of it this way, or around "older than creation" which is just a way you can think of a thought. This is exactly like science. √ There are probably skeptics that will not like this, book, and what it is about. These are probably, lost people, yet I feel that the world is about the oyster that you can open, to bring up anything to the table. This is bread and wine. √ This is earth, water, sky, wind, fire, and assorted various lies, that do not come true. Yet, I am affirmative that my theory, is true, in most circles, and in most people's opinions, or I would not have written. This is about deformation, and naturalization. These are about the books, that I have chosen, to write, and about and with this, I have chosen, to play, and work, and about this, I have chosen, to also accompany people in their life and works, equally. This is with what I have written. These are all about the allegory, of the doctor and the patient. These are about how a doctor, and a patient can be healed, with magic. This book heals with fact. This is a strong reassurance that fact, can set a person free, and opinion cannot. These are about the two realms of factual existence, and about how the realms, of thinking and formation, of opinion, is about the relation, with God, and man, and about how, God and man, can overcome, these theories, of God and man, and about how, the difference of God and man, cannot be about the triumph of man, and how he is about the losses of the Garden of Eden, which only happens, if you are insane.

This is about the evil of the world. This is about how the evil, of the world, is about the triumph in the Lord. These are about the agents of things that are about freedom. This is not just with naturalization, but with also, the theories of man, and how he goes to hell. This is only if he chooses this to be. This is about the ways that acting and choosing to be acting. These are the agents of desire, and conformity, that are act, and ready, for issues, of environmental protection, and the "b. s." of our universe. I am a real person, with real thoughts and ideas. These are what I express. √ I am not a natural phenomenon in writing, but more or less, an intellectual theorist of the Christian faith, and practice. √ This is about the way the world works that is about the ways the world works. √ This is about the way the world works that is about the way the world works that is about the way you and me work. We work the way, we are programmed, to be, that is about the nature of the human mongoose, to the snake, in my philosophy of family fun. This is a symbolic meaning, that occupies my thought, in the realms of thinking. These are thinking, that are about the philosophy, of reason that is about the ways that the philosophy, is about the ways that are about the mankind tales of loss and gain. This is about a snake ring that is about the bond, between our family, that is about the contest. This is survival of the fittest somehow, that has contested for many generations, without a relation to the reality, of man, and nature, and ways of philosophy, that are about the contesting of the natures of man, and about how the natures of man, are about the natural philosophy, that is about the "Benjie the Hunted" movie also. Yet, Darwin wasn't the one who created this, he just wrote it. He is the author of this though. This is where my comparison with Jesus Christ, comes into play. This is about the forgiveness, that is about the creation, of

the world, that is about the ownership of the words he said. This is about the philosophy, of the greatest philosopher in the world. This is about the ways, that philosophy trained him, of other people's that is not there, and is not prevalent, and is not associated, with the order of thinking, that is about man. √ These are the issues, of society, that Jesus Christ, created, and associated with Himself, as well as everybody, in the world, that knows Him. He is the Word, which Darwin, was not. So, why do we contest, if evolution is true? This is purely because of school. If I were to say, that different people, were different elements, then you would say I was a lunatic. This is because of Jesus Christ, who believed in that He was the church body. This is about the ways, that the purest and most simple, ways of man, that are about association, of many different people, with many different things, occupied in the reality of the world. This is through space and time, in the elements, and the creation of the entire world, through elements, that are because of the materials. These are, of and wholly are about the occupation of time in circumstance, which is a, "would you believe me", that was true, wholly, about the creation of the world. The theory, that I have is that 1. Different people are different elements, and 2. The creation of the world are through different elements, and 3. The difference between different elements, are about the nature and the concern of what created them, and why they are there is because of different people, and 4. The different, people, who know this, are not lunatics, and 5. The different people who believe in this, are naturally this way. My creation of the world theory, is about the different theories and aspects, of the world way, that is about "becoming" that is about the nature, and the philosophy, behind, the elements, that are about the ways that man, can "become" whatever he be-

lieves in. These are about the philosophy of the elements, that are about the "make up" of the elements, that are about the belief theory of the world, and how it got created, from the "Word of God", except for the "Serpent". The theory of how Satan, came around is, according to me, a hypocritical statement, of how the world evolves, from the elements, of desire and influence, that are about the science and math, of the different ways, that man achieves and maintains his "weirdness", of the certain philosophy, that he does not practice, but is forced into believing, about the materials, and how the materials, are occupying time, and space, and are about how the world, is created, and maintained, is through the certain context and about the science and math and about what the science and math truly mean? The certain meaning, of this, is about and for, the context and content, of and about the real realm of the future and about how the future is about the way man, is about the way, the creation of the world, is about the materials of the world, combine and create, the materials of faith and justice, and why we even think the way we do, that the world was created, by the future elements, of man's desire, that are purely fictional. These are about the ways that man is created and about how the man, and the nature of the world, became, with the nature of the world, is about the creation, and dominance of the world, was about and is about, the nature, of the world, and how it was created, and became, through justice, and harmony, in the natural order, of justice, and the natural way justice, became which is about, the ways that justice, became and becomes, and is about how it is, and about how it becomes, much stronger, and more reliant, on people, who donate and respect, their time, on their circumstance, which is about the circumstantial happenings of the world, and about how about, if we believe in this,

there is a notion, of justice, and about how justice, is about the research, of the different worlds, that are about harmony, and dedication, that is about what, which is what we are and is from what we are about in whole, and what we are about in this which is justice: 1. This is and are about how we can justify, the way the world works, and 2. how we can exemplify what justice means, through the sanctification, of what the world means, and 3. that is about the ways that what makes us feel, this, is pure insanity, and 4. What this is, that is blamed on Jesus Christ, is what is for not forgiving us, which is about for our sins, to be forgiven, and exemplified, in Christian thinking, that 5. What this is about is about forgiveness, and how this is blamed, on us, for not believing in Jesus Christ, and 6. This is about how we believe in things. These are programmed in us, from day one. The fact, that programming, the way and the fact, that mankind, and man, exemplify the teachings of one person to another are about blame and concerning one another, with this, is about the memory, and reflexology, that is about the nature, of man, and the mankind, that is about the tales of gain and loss. These are also, about, the nature, of man, and mankind, and God, and the real, Jesus Christ, that is and are about the nature, of man, and mankind. This is about the whole philosophy that I have invented.

My Theory of The Origin of the Universe-

The theory of mine, is about the lying, and cheating, and stealing, of other's. These are about the reflexology, of modern man, and how the nature of reflexology, are about the ways, that the nature of man, is about the nature of

God, and about how the nature of man, and God, are always interrelated. These are through the elements, of earth, fire, wind, water, sky, and air, and about how we must survive because of this, factor of communication, that is in our world, through the belief, of the natural world, that is invested in man.√ The fact, that man, is that what is about what he is about, is about naturalization. This is about the ways that naturalization, is about the nature of man, and about the naturalization, that is about man. This is about mankind. This is what man, is about, that is about mankind. He is about the ways, that mankind, is about man, that is about mankind. These are about mankind, that are about the way, mankind, that is about the way, man is about the way, that is man being about mankind, that is about explanations. This is about survival of the fittest. The way, that man, is about, survival of the fittest, is about the ways, that mankind, is about man, that is about mankind, that are about mankind, that are about man, that is about naturalization, that is about mankind, that is about the real mankind, that is about man. √ This is about man, that is about mankind. This is about mankind, that is about mankind, that is about the mankind. He is about mankind. These are about mankind, that are about mankind, that is about mankind, that is about mankind. .√ This is about naturalization. These are about mankind, and how he is about mankind, is about mankind, and this is an explanation from naturalization. All people, can realize and understand mankind. This is about the understanding, that is about mankind. These are about mankind, and what mankind, is about, is about mankind. .√ These are about mankind, that are about mankind, that are about mankind, that are about mankind, that is about man, and mankind. These are about the games of life. This is about the importance, of cooper-

ation, and group work. These are the essentials of success, and the failure, and the success. This is how we realize the life. This is about life, and how the life, is Jesus Christ. .√ These are the tall tale societies, that are about false, and strange notions, that were not about God, and Godliness, but about Godlessness, and about how Godliness, and Godlessness are about the human being. .√ This is about the way, that naturalization, is about the client, and student. We are naturalization, in school. .√ These are about the ways, that mankind, and naturalization, are about the society and the world. .√ These are what the society, and the world, rules for, and of, which are about naturalization, and the context, and how the context, is about the control, and context, which is of and about naturalization. These are about blame, and bother, and how we are completely innocent. This is when we study naturalization. This is the cause of naturalization. We are about the cause, of naturalization, that is about the cause, of the cause, that is about the naturalization, that is about the cause, that is about naturalization, that is about naturalization, that is about naturalization, that is about the naturalization, that is about the cause of naturalization, that is about the whole, and the part, that is about and for, naturalization, that is about naturalization, and about how the naturalization, of the world, is about nature, and man, and the God, and about how the naturalization, is about the nature, and man, and God, and mankind, and environment, and also naturalization itself. This is through fire, water, earth, sky, wind, air, and the elements, that make up our universe. These are about the colossal ways, that these elements make up our society, and the way we view our outside world. This is through the dynamics, of creationism, and how the earth, was created, through the content, of the contextualization, that is about the nature,

of the earth, and how about, the nature, of the earth, is about the content, and context, and that is about the control. This is over the formation, of our outer world, through our intellects. This makes sense, because we are not the earth, as the people, who make it up. Rather, the earth, is made up of elements, and anyone, who is crazy enough to think, that these are people, are very strange. These are elements, that are about the people's perception, that is about the way, that the people, are and view these elements. These elements, are the view, of the natural, ways, of man. If someone became an element, then we would not survive. So, people are definitely, creators themselves, and not just what they read. We are not a horoscope. We are not an animal in the Bible. We are not the God, that we worship. But, naturally, we can get used, to, the land that we live in. This is through learning, through the elements. These are what could have shaped us even. This is through our perspectives. .√ This is about the ways, that mankind, and man, and the mankind ways of man, are about the elements, and how we think in terms of these. This is primarily through our health. The elements, that are through these, could be the reason why, we do not "evolve" but change into the newer "us". If you drink more water, naturally, you become more prone to be, healthy. And, if you see the sky, you are sometimes happier. Also, if you play with fire, you might get burnt. These are about the ways, that elements have shaped us, and our perspective. This is why. .√

SATAN AND HIS ELEMENTS-

The world is about the essence, and is about the essences, that are about the new, and old, ways of naturalization.

These are about and in, the context, of reason, and rhyme, and how this is about learning, is about the learned. We can learn as these elements. These are what we have to have, to survive. This is air, to breathe, earth, to live on, water to satisfy, fire, to learn and help keep us warm. We also must have wind to make a difference, and these are the six elements, that do make up, our world. These are what they were created from, if there was a man that we came from. He is the real Jesus Christ, and He performed miracles in these, elements. To nature, and man, these are naturally, prone to survive with him, as his elements. These are also, learned, ways, of naturalization, and about how the naturalization, of man, is about the control, and learned, ways, of mankind. .√ This is about the ways, that the whole world, is about the new elements, that we always go to. .√ The ways that alchemy work, are not these. These are the nutritional value chart. These are known as the "survival elements". These are not weak and strong elements, but the real elements, that earth, must have been, with, before we fell from the Garden of Eden. .√ This is about the way that the earth, was already this way. These were the elements, that God created the earth with. This is about the ways that man and woman were also about these. These were about the ways that are about the ways that nature and man, and God, come together, with. These are what symbolize, the right and wrong, ways that we do things. These elements do not go away. They just stick around, and make the world the way it is for man. .√ Without these, there would be no earth, and no sky, and no wind, and no water, and no fire and no air, and no elements that are about the earth. I think that the world did not evolve this way, but was created this way, in nature, and man. These are also about how God was created. These were from the fall of man, and his "Ser-

pent" who must have made this, this way too. This is not impossible. After, man fell because of "knowledge" and sin, then there was these elements. Maybe, Adam knew the world, this way, already. But, there is yet to factor in time. These are about the ways, that time, could be a real thing. This is not just theory. .√ This is about the way that the world is about the way the entire world is about the formation of the world that is about the excellence in how we view it. This is through the ways that man and mankind and God and nature and environment, shape us, and how we know all of this is through God. He is about the ways, that the ways, that affect the ways, that affect the ways, really do affect the ways, that nature and man, are a part of the natures of man, and the nature of man. This is what we like to call harmony. This is how this works together. These are not stupid theories, but are the theories that are about man, and about God, and about nature, really are about the truth that sets us free. This is about the nature, of mankind, and how the nature of man, is always affected, by the God of the world. He is in Heaven on the right hand side of the throne. This is about the ways, that nature, and man, are about, and are in, the society, of man. These are what make the rules, no matter what the cause of the world's creation is. These are the church, the school, and the government. These are about the ways, that man, and environment, can persuade, and knowingly, know about the knowledge, that is about the knowledge. This is through the elements, and how these are about our thinking also. These are about the ways, that are about the ways, that are about the ways, that are about equality, and the essences, of man. We have to follow something. And, this is the ticket. This is about the ways that the naturalization, is about the causes, that are about the cause, that is, about are about the ways, that mankind, shares

with himself. These are the knowledge, of man. These are the strength, of mankind. These are the creation of God. These are about how we rule, together. These are from, the essences, of man, that is about the elements, that are about the body of the world, and how we are made up with it. These are the elements, that are about the elements, that are about the elements. These are about the way that the nature of man and God, could be made from these elements. These are the allegory of the doctor and patient. This is that health, is what makes us living, and to have health, we must go to a doctor, and have it checked. These are the ways that man, and his nature, are about the God, and his nature, that are about the mankind, ways of thinking, that are about the colossal. These are about the ways, that man and God are about the content, and context, of the nature, of man, and the God who created us. These are in whole. These are the parts, of justice, that are about the rules of our society, that are about the message, of the Word of God, and about how we, known as man, and the whole of us, known as mankind, is and are about the beliefs, that keep us strong. People who go, against these, beliefs, are considered "abnormal". These are the "smartest in the world" kind, of people, that believe in everything, and then], from that, know it all. These are also naturalization fans. These are the fans, that must be, compared, to the world, that is about fact. These are about the content, and context, that are about people, that are about people, that are about people. These are people, who are about the future, who are about the past, that are about the present. These are the containing knowledge, people, who are about the containing knowledge, that are about the containing knowledge. This is about God. He is the one who makes us believe in Him, who knows our life, and who knows, our promise. He is

the one who keeps the promise. This is about the knowing that is about the knowledge, that is about the theory. This is about God. This is because I am a theoretical Christian, and how we learn, is through the dynamics, of the faith, and works, of God. These are what save us. These are the looks, and are about the time, that is about the colossal. These are about the understanding that is about naturalization, and how it is not hypocritical. This is the best, and most smart, theory, as well as the most powerful, that is about the context. These are about the colossal, that is about the concern, that is about the control, that is about the ways, that are about the sinning that are about the creation, that is about deformation, and how to master this. This is about the ways, that master and slave are about the control, and no one who knows where this comes from. This is about the sky, and the water, and no one who knows how it formed, or how it originally got there. This is about fire, and why it is about what we are concerned with, sometimes. These are all controlled, by earth that created all of these things. These are about the ways, that client, and teacher, relationships, do last. And, this is why and how these people, are the reason, that we do exist, and do this to each other. These are the ways, of controversy, and are about the ways, that controversy, is about the way that man and mankind, are about the ways, that we need to know and must know how to succeed. These are about failures, and about how we fail, and succeed. These are with the way, that elements, also survive with us. We are comprised of mostly water. We are always about breathing, and challenges, of lung. We are also easily burnt by fire. These are about the elements, that are about the world way of being created. And, thought comes from evolution. This could be the biggest truth, or lie, in the history of the world. Yet, Darwin, still is in

schools. This explains that thought must come from evolution, if we believe, that we are monkeys. Also, that we must compete. In naturalization, thought comes, from you, and you alone. The theory of thought is about the nature of man and about how he goes along with the world, and the ways, that the world are about the control, that is about the ways that the control, over the world, is what we believe in. There must be some way that we believe in time, and how we are the most important, ways, that we are living, and how we are living, is about the ways that are about the ways that are about the substance, and essence, is about the context, that is about the content, that is about the ways that the conflict, and interest, are about the ways, that are about the control, that is about the controlling, and the ways, that we are controlling ourselves, is through the elements, of desire, and how we were "tempted" to believe in any of this. This is through naturalization. This is about naturalization, that is about the earth, and the elements, of the earth, that are about the control, and harmony, of its people. This is about what we are about and what we do, and is about the ways that the elements, are about the colossal, ways, that are about the time, and space, which are about the ways, that are about time, and space, that is about the time, and space, and fluctuation, of time and space, that really is created, by our own knowledge, also. We must have fell. We must have sinned. And, we must have created a lie, about ourselves, slowly. These are the created notions, of God, that are about the created notions, of living, that are about the creation, and the control, of the future, and with time, the space and time, that we all occupy. These are about the elements. These are what created the world. But, these are with God. God spoke, and it was created, the way He wanted it to be. God wanted the earth to be

this way, or it would not have been this way, in whole. This is about the way the theory of God is naturalization, and should not be compared to other theories of this essence, and quality. The theory of the explanation, of God, is about the essence and substance, of how God exists. He exists through quality and is about the quality of the essence, that is about the quality that are and is about the ways that man and woman, have been created. They are about the quality, that are about the quality, that are about the quality. These are about the quality. These ways that are about the content, of the context, is about the control and context, that is about the ways that man increases and decreases his quality. He is about success. These are about the essence and quality that are about the control, and the construction, of this control, that is about desire. For some reason, we like to do things. This is an explanation of the greatest form of knowing, ever, known as desire. This is about the way the world is about the way the world is about the way the world is about the way the world is about the way the world is about the way the world is about the way the world is about the way the world is about the way the world is about the world. This is about the way the world is, about the world. This is about the world that is about the way the world is about the world, that is about the way the world is about the way the world is about the way the world is. This is about the way the world is about the way the world is about the way, This is about the real Jesus Christ. This is about what the real Jesus is about that is about the way, the world, is about the way, the world is about the way the world is about the way the world is about the way the world is about the way the world is about the way the world is about the way the world is not about power, but about authority. This is what rules, the world. This is about the way the world, is ruled by the world. An

authoritative figure that is about the world, is about the content of the world, and the way the world is about the God, of the world that is about the God of the world. This is about the God of the world and how He is about this. He is about the context, and content, of the world, that is about the content and context, in the world. This is about the way the world, is about the way the entire world, is about the way the entire world, is about the judge. This is about the ways, that we judge things, that are about the stupidity, that is about all of us. This is through the sanity , of the world, that dictates this. This is about the ways that the entire world, is about the collective, and generalized, ways that sanity rules it. This is with the theory of naturalization. X The way that the world is about the way the world is about the world, is through the dichotomy of good, and evil, and how these two work together. These are about the ways, that the world, are about the collective, that are about the corrupt and strange. X These are about the strange and unusual ways, that people, are collectively, about and from, the dichotomy of good and evil. These are about the good, and evil, and the ways that the world is about the ways that the world is about the complete and complex. These are usually considered "smart", but are really sane, in my opinion. This is about the ways that the world are about the sanity and not just the "smarts" because of strange people. These are the people who think that they can get away with anything. They think that Jesus does not rule the world, but think that the world is ruled for and about the ways that direction is headed, in the line of destruction. These are about the ways, that the construct, is about the construction, that is about the collective. These are about the collective, and generalized judgment, that is about the ways that the smart are not about the sane but the sane is about the

control, of the sane, like the smart, is the control of the smart. These are about the smart ways that are about the smart ways, that are about the smart, ways, that are about the judgment, that is about the way that the entire world, is about the way the entire world is about the way that the world is about the way the world is about the way the entire collective, is about the premise, and judgment, that are about the control. These are about the entire worlds of people, who are about entire worlds of civilization, that is about the content and complex. These are about the control, that are about the ways the entire world are about the judgment. This is about the ways that the world is about the ways, that the entire world is about the God. X These are about the ways that the world is about the ways that the entire world, is about and from, the essences, that are about the essences, that are about the essences, that are about the control, and authority, that are about the conflict, and construction, of the world. These are about the way that corrupt people, do not rule the world, or else they are "sick" with power. This is mainly authority, and the authoritative methods of science, and math, and languages that do not make sense in school. These are about the completely understood ways of man, and about the way man is about the government, is about the collective. These are many people, designed, to do the same. They are in the real governmental systems of our social standards and essences of knowledge and knowing. These are about the collective, and how the collective, is about the collective, is about the collective. These are about George W. Bush, and my favorite thing. This is the Presidency. This is about the way the world is about the ways that the world is about the way the entire world is about the content that is about the containing methods that are about the ways that the world is about the ways that are about

God. This is about the ways that the numbers and math are about the science, and about how the math, and science are about the numbers. These numbers are about the ways that the numbers, are about the ways that are about the ways, that are about the ways that are about the ways, that are about the ways, that are about the ways, are about the ways, that we do business. This is about the common world, and how it does not have any sense. These are about the ways, that the entire world, are about the concern for others, and how this is gone. These are about how the entire world, is up to us, and how we follow it, is through the grapevine. This is why we must be sane, and smart. These grapevine rumors start only evil. These are also considered, also, evil, and good, sometimes, when they are lies. These are about the way the Presidency, doesn't do this, and the way the Presidency does this, is through denial. We are all the same, and affect the ways that affect the ways that affect, the good and evil. This is through touch football. The ways of the game, are about how we play it, and not how it merely looks, to others, in these same circles. These are about the carefree and fancy free, ways that the notion of the God, is about the notion of man. These are what the world, is about and what this is about, is about the essences, and concerns with man. These are about man, and how man, and God, are about terms that we do not understand together, These are about the ways that man, and God, are about the ways, that the entire man, and the God, work together. God created, words, that are as smart as Him. This is unless you are against the world. These are the words God made up. These are about the rules of the game. These are compact, and complete, words that affect you. These are 1. Love, 2. Joy, 3. Peace, 4. God, 5. Patience, 6. Kindness, and 7. Glory, and 8. Many more words, that affect you and me. The

words of the world, are good and practical, if they are God's own. These are about the colossal, ways of understanding, that God Himself made, for us, and for each of us, in His own terms, and His own language. Yet, we do not understand them. This is unless we are sane. These are the issues of modern man, that are about the issues, and are about the complete, and complementary lives, that God gives, to people who love Him. This is in the same way, He loved us, for others, and the same way that God loved us, for Him. These are about the ways, that the man, and God, are about the conclusion, that we draw, from others, and about how the conclusion is not right, usually. This is because we forget, and press on. X These are the conclusions, that last, and that will last, and that will occupy, time and space, within us, forever, and forever, and forever. These are these words, and how they result, is through triumph. At, the end of the race, we are about, the ways, that we are challenged, and about how we are challenged, is about the ways, the direction, is about the ways, that the entire world, is about the conclusion. This is about the control, and content, that is about the context, that is about knowing, that is about complete and total, judgment, that is about other people, and about God. Yet, God judges, us, and we do not judge God. He is in many people, and do not judge them, but they try to judge themselves, which is very confusing and strange. This is how the world has adapted. Because of this, we have doctors now, who place emphasis, on sanity, and sane notions. This is so we do not evolve backwards? No, this is because, we all know how to treat, one another, and how to compromise, when situations, get tough. These are about the ways, that the notion, of God, and Jesus Christ, are about the content, and context, of becoming, that are about the same, as the race, of God, and of man. God be-

came a man, to be able to judge us, and live with us, and know us. These are about the content, and context, that is about the content, that is about the context, that are about the content, that is about the context, that is about the world, that is about the world, that is about the world. These are common resentments. These control our God, and we control Him. He is the one, who is judged, by God, and He is the one, created by God, to be Him. This is known as "God's Son". This is about "God's Son", and is about the ways that God's Son, forgives. He is about the ways that God, is about, the ways that God, is about, the triumph, that is about the glory, that is about the honor, that is about the majesty, that is about the honor, that is about the decisions, that is about us, all, and all of us, are about all of us, in whole and in complete, and total harmony. This is about the ways that the glory, and honor, and majesty, are the ways He judges us. This is to be, wholehearted, and complete. These are about the ways, that the entire, control, of the world, is about the Book of Life. These are where the names, of saved, Christians, are inscribed, forever. X These are about the ways, that life, is about the honor, and glory, and majesty, of God's, that He is for and about, in time and space. He is the one who already judges, us and makes it the rest of our life. This is about the ways that the world, makes up the world, that makes up the world, that is about the ways, that the entire world, makes up, the entire world. These are about the ways that are about the content. These are about the content, that is about the content, that is about the content, that is about the content, that is about the content, that is about the great and mighty, God, that is about the good, that is about the goodness, that is about the judgment, that is about the control, and the weird and strange, people, that are about the content, that is about the context.

Riley's Natural Naturalization

This is about the Book of Life. This is about where the names, are inscribed, and taken down, by the Lamb, known as Jesus Christ. He is about the ones, who are inscribed, in the Book of Life, where these people are saved. These are about, the ways, in which are about the, content, and context, of the beginning, and the end, and where we end, up, is the essence, of man, and about how he judges. This is about how Jesus Christ, the Son of Man, judges, that makes up the whole world. These are about, the whole world, that is about the content, and complexities, that the life of this throws at us. These are about the tools, of the trade, and not touch or flag football. These are the life skills that we all learn, and are all about. These are the life. These are the way, and these are the truth. These are only, of Jesus. These are about, Jesus Christ, and about how, the real Jesus Christ, is about the know, is about the entire world. This is an act. What we know about Christ, is what we know about Christ. And, what we know about Satan, is what we know about Satan. Opinions, are what we know about opinions, and that is it. This is about knowledge, that is about knowledge, that we all fell with. This means that I am not better, or worse than you, but we are equal. Knowledge is about knowledge, itself. The knowledge is about the knowledge, that is about knowledge. This is about the knowledge, that is about the knowledge. If you are an opinionate person, then you cannot go to hell, or heaven, unless you receive the free gift of Jesus Christ. This is about the real Satan, and how he rules the world. The knowledge, of the real Satan, is about the real Satan's knowledge. Yet, he is a lost cause in the Bible. So, if you are a Christian, then you are forgiven. This is about God's immaculate conception plan, and how it rules the world. This is about the world, and how the smarts, of the world are about the power of

books. These power of books, are about the essence, and substances, of the real power, of the real world. We learn about the powers, of the books, by the way books, are arranged. This is about the real, way, that books are about the books, that are about the books. These are about the books. The power, of books, are the explained, to people, and the ways that people communicate with them. They are about the power, that is about the power, that is about the power, of people, talking to other people, as a lecture. This is about power, that is about power, that is about power, that is about the explanation of the power behind the books, that are by the power that cannot explain everything. These are books, that are books, that are books. These are about the power of deformation, that is about the explained, and the theory behind it that is sane. This is true or else I would not know what I was talking about, in the real existence, of a real discovered new art form. This is about the explained and how we cannot explain anything to anyone, usually without the precise. This is about the brand new theory and how the new theory is about the, explained, that is about the explained, and how this is about knowledge, is about the theories of knowledge, and how they trust, and are about trust, from people. This is about the ways to a brand new agenda. The fact, is about the fact, that is about the fact, that is about the fact. This is about the fact, that intelligence, ruled the world, and still does rule the world, of the discovered, people, who know about the world, that know about them ruling the world. This is from the perspective, of history. This is about history, and about how history, and about how the history, of the world, is about the knowing, and knowledge, that is about and are about the world. This is about the usual world, and how there is a religion, behind everything. These are elements, that are

about the elements, that are about the elements, that are about the not structured, figures, of a language, that is beyond words, and meanings. The meanings, beyond, the meanings, are beyond the meanings, that are beyond the reasons, that are beyond the reasons, that are beyond the reasons. These are beyond the reasons. This is obvious how I am writing, a theory, that is about acceptance, and how the acceptance, is about the context, and content, of the world, of writing, and how the writing is. This is about the context, that is about the content, that is about the ways, that are about the written structure, of God, and Jesus Christ, that are about the philosophy, of the science, that is behind the science, that is about and meaningful. This is about my passion about the unexplained and about the way that the world, is about the meaning, of the world, through the meaning. The explanation, of meaning, is about the ways that beyond, the meaning, is beyond the meaning, that is about and beyond the meaning, of the world. This is about the way the world is about the way the world is about the way the world is about the way the world is about the way the world is about the way the world is about the context, and content, of reality. This is about the ways, that natural realities, are about the context. These are about the content, that is about the research, that is of the ways, that the natural, ways that we behave, are about the philosophy. This is what the "love of wisdom" is. This is about the ways, that Satan is defeated, and we all know the world, better than other things, like repeating the words, that are about the philosophies, of nature, and man, that is about the nature, and man, that is about the nature, and man, that is about the nature, and man, that is about the man, and the nature, that is about the nature and God, that is about mankind. These are about man. This is about the ways, that

the man, and the ways, are about the God and man, and about how the God, and man, are about the essence, and is about the substance, that is about God, and man, that is about the God, and man, that is about God and man. These are about God and man that is about God and man. This is about the way naturalization works. This is about and is through the entire world, which is controlled by royalty, known as the powerful people, as well as Presidents, and the Senate and the Congress of all different countries. These are about the ways, that the entire world, does things, and is about the control coming from the government. The kings of countries in the government, are the royal people, who do rule because of God. The Presidents are elected because of God. And, the government in non-corrupt countries is controlled by God, which is the greatest pleasure. This is to be in the government. This is about the naturalization of the government. This is to be a hard to get job, and a higher position, than God, but only on earth. Jesus was a carpenter, and He knew, about all the governments, because He hired them all, rightfully. This is unless they went against His will. The church and school, are also controlled. This is by the world, that is controlled by the people that is controlled by the state that is controlled by the united people, who are controlled by the formation of the schools that originally formed. This is how the United States formed, through church persecution. These are about, the ways that the money, are about the ways the system, are controlled, by the people. These are about how the people, are about the people, who are always about the government, that are always about the United States, that are always about the Senate, the Congress, and the people, who make this up. Also, in different countries the people, are the same, yet, they are controlled by the body of the

government, and the education, and the work force, that is always, about the church. The church is the one that is most often, thought of. The school is the next. And, then the government, is controlled, by the entire world stage, of everybody. This is in the modern day, Heaven. In the past, it could have been, hell. These are the separations of war and peace. If Christ, is the King of Peace, then the government, is the body that decides on this too. This is why peace prize winners, are heralded as so important, by me. This is because it comes directly, from peace, which is the churches. The way peace, is offered, by the people, of the public, that are and is about the direct, and control, of the direct, and total opposition, is about parties and about how we have leaders. These are the leaders, of the world, that have parties. These are associated by them. When a leader, doesn't have followers, then he is considered evil. This is because he always goes to the weak. These are without others, to, who lead. These are the coordinates in naturalization, with the equation. It is naturalization, that explains the way, that the world is. This is always an explanation. This is about power. The power, in the ways, that naturalization, is explained, is in and about, the equation, that is in and is about, the context, that is about the equation, and about how the equation, is about the equation, that is about the equation, that is about the equation. This is about the equation. This is about, the way, the world, is about the sense, and senses, that are about the essences, and the essences, that are about, the way, that the direction, follows and dictates, the circumference of a circle, and about how, the entire, world, is made up of mathematics. These are about the ways, that math, and science, are about the progression, that is about the body of Christ, that is about the sense, that there must be elements, that are designed and placated through the entire

world. These are about the elements, that are about the difference, that is about the passage, that is about the rights, and reasons, that are about the circumference of a circle, and how naturalization, is related to that. This is about the direction, that is about the ways, that are about the ways, that are about the direction, that are about the place, that are about where we are headed, that are about the circle, that are about the control, that are about the, dynamics, that are making sense. This is in our world. The way, Satan, makes it, in our world, is through evil. He runs things through evil. And, evil men, help out, and the control, from the evil is about the context, that is in evil. This is about the way that evil, does things, that is about the way the evil, exists, that is about the way, the evil, takes shape. This is about the coordinates, of peace and war, and how they are confused. There is always peace, unless there is a war. And, the elemental, way, of describing this, is through the periods of thought, that we go through, and experience. These are the experiments, that are about the ways, that are about the ways, that are about the ways, that are about the progress, that are about the client. This is about the relationship, that is about the progress, that is about the progression, that is about the way, that is about the way, that is about the way, that is about the way, that is about the way, that is about the way, that is about the evil. This is about, an evil, way. There is evil, in the form, of evil, that is about the forms, of evil, and how we experience, these phenomena. There is a plan, that is about a plan, that is about justice, that is about the ways, that are about justice, that are about the gain, and the loss, and about how we are a part, of the gain and loss, that is about the justice, and about how the justice, is about the losses, that are about the Godly, that are about the direction, that are about the

sense. These are about the senses, that are about governing bodies, that are about the daily routine. These are about the client, and the ways, that the clients, are about the help, that is about the help, that is about the help, that is about the helping. These are about the client, and is about the progress, that is about the client, that is about the way, that is about the way, that is about the way, that is about the way. This is always about, Jesus Christ. He is not a liar, but a lunatic, which is not a lunatic, but by some strange opinion, a false teacher? No.

Jesus is about the concern that is about the relationship, that is about the ways, that is and are about the client, that is about the real student. This is like this, in many ways. This is about the ways, that God, and man, are about the philosophy, that are about the client, that are about the progress, that is about the client, that is about the opinion, that is about the sense, of the senses, that are about the senses.√ This is about the ways, that natural, ways, of progress, are about the progress, and about the ways, that progress, are about the ways, that progress, are about the way, the progresses, are about the conscriptions, that are about the knowledge, that is inscribed in the Book of Life. This is about the way, the entire world, is about Jesus Christ. He is about the superstar. He is the man, who invented, everything, because of Him. And, yes, He understands us. This is in whole. The whole explanation, of what is the cause of the universe, is about the cause of the universe, and how the whole explanation, does not need apprehension. The conclusion, with this is, that the universe does as it does, and no one can stop this, from happening, including Jesus, because of God's promise. This is to never end creation. This is about the promises, and how they are kept. This is between God, and Jesus. He

came to the earth, to solve, man's dilemma. This is about, the participation, between good and evil. There is always, a reason, why, the world is, as it is, and this is through God's creation. There have been, prophesies, that are about Christ's death, and the prophesies, about Christ's coming, and prophesies, about the coming of the age of God, which is after Jesus came. This is how the real world, functions, and is about how the world, does good. This is through Jesus Christ, only. This is about the prophesies, that are about, the conclusion, that are about Jesus Christ, and about how the Trinity, is about the sense, of man, and mankind, and how there is a second coming. This is of Jesus Christ, in the world, and about how, the earth, will be baptized. This is where no explanation, will be needed, and the Christ will be seen, as a pure Lamb. This is of a Lamb of God. He is about the conflict, that is about the ways, that man, and mankind, are about man, and mankind. This is through Jesus Christ.

Satan and Naturalization-

There is a ripe sense, of becoming. This is in the evil, of the evil, of the evil. This is of Christ, and Christ's second coming. There is always a angel, trying to win. This is the only explanation of Satan.

He tries to be an angel, of light. Then, he tries to be a dark angel. There is a common definition, of Satan, that is of and about the clear, and present, that is about and because, of, the sense, of rescue, and opinion, that is about and of, Jesus Christ. Satan, likes to manipulate, people, and cause lies, and deceit. This is to his followers. These are the Satanic cults of our nation. This is about how Sa-

tan, rules over these. He is a false God. And, he is the resurrection of evil. He can walk the earth.

NATURALIZATION AND SATAN-

The naturalization, and Satan, does not exist together. This is when naturalization, is treated realistically. The real Satan, is about the judge, that is about the strength, that is about the courage, that is not about school. Naturalization, is in school, and is about, the belief, that common things, do go together. Yet, Satan has to do with other things.

NATURALIZATION-

The essence, and appeal, of naturalization, is about the essence, and appeal, of nature, and man, and naturalization, and mankind, and God, and environment, that is about the elements. These are the elements that formed earth, to be this way, toward God and man. These are about the ways, the man, and his naturalization, is about the ways, that naturalization, and man, are combined together, and sense, a result, among, the earth, and its elements. There are elements, that are about the elements, that are about the elements, that are about the elements, that are about the elements, that are about the elements, that are about the elements, that are about the elements. Some, people, want to have the elements, of the earth, to themselves, but the earth, is dispersed, this way. There are no more elements, that there are. There are only six. This is how the elements created the earth. This was with Jesus, and God. Yet, there is a science, and a math, that is about the ways, that the elements, also, touch and go with

each other. The elements are about the ways, that are about the way, that the elements are about the God, of the world, and about how, the earth, is about the earth, that is about the earth. The elements of the earth, are consistent with each other. They are made up to be one another. They all come together as earth. They are the ones I have described. Water, air, wind, earth, fire, and sky. These are the elements, that are about the elements, that is about the ways, the elements, are about the way the elements, are about the ways, the elements, that are about the elements, that are about the elements, that are about the elements, that are about the elements, that are about God, and Jesus, and man. These are what Jesus had, to have. This is what God created, this way, and these are the elements, that created the earth. These are the elements, that made up the earth, that are about the consistence, of elements, that are about the evolution, of the earth, with its elements, and how they are about, the earth, and how there was no evolution. There was a theory, on one, that didn't work, and looked strange. These are the elements, that are about, the elements, that are about the elements, that are about the elements, that are about the strange, and weird, ways, that man, and mankind, are about the elements, and how they continue. They will always continue. The riddle is that if the elements, were created, and we are part of Christ's body, then how will we be created. This is before the creation of the earth. This is a question, in science. This is about how the ways, that the earth, and the elements, are about fire, water, air, wind, and sky, and earth, how in all consistency, must the earth return to its normal place? This is through knowledge. The knowledge of our outer world, are what makes us very smart, in the ways of insanity. This is about believing lies, that the whole earth, was created by one voice. But, it

says "Ourselves" and people lie about this all the time. The earth, and its containments, are made from the earth, and the way it has changed, and turned into something new, always. And, in return, we have become smart. This is with the ways of the earth. The position of the earth, is aligned with the stars and the atmosphere. This is about the way the earth, is about the ways, that the earth, is involved in the circumference, of the nature, of man, and his alignment, with the universe, and the elements. The elements are what create the earth's pull, and substance and essence, with the elements. This is how naturalization also works. This is through the sky, that God created, the wind, the fire, the earth, and the water, and air. This is about the coordinates, that God has charted. This is how the voice formed, and became. This is through the body of Jesus Christ, when He was with God, in the beginning. He is God's Word, so He was with God, in the beginning, when God spoke. This was with a voice, that created all of this. The voice, that was spoken, was hovering over the waters. This was with the spirit of God, in the beginning. Has God always been around? No. He was also created when Jesus Christ, was created. This is through faith. But, this is one hundred percent better, than Darwin, who thought we were created by an explosion. I think that God's voice, was part of the Spirit that came from the elements, that came from the creation of the world, with God's voice. This makes it easier to understand creation. These are the created outcomes of the world. These are the spirit, the Word, and the Naturalization, that has a explanation, for all things. This is about the explanation, that is about the God, that is about the earth, that is about the sky, that is about the air, that is about the water, that is about the fire, that is about the wind, that would not make sense, unless you studied the theory of God. This is naturaliza-

tion, in its truest format, and thinking, and numbers, and language. This is a pure understanding, of God, and man, and nature, and the nature of man, and the nature of God, and the study of environment, and the ways naturalization, itself is in the equation. This is also with, mankind, and also with the logic. This is about the logic of creation. These are the logics, of creation, that are about, the creation, that are about the creation, that are about the creation, that are about the creation, in study. There is a myriad of points in the nature. This is of Jesus Christ's that was God's that was man's, that was God and man. God, and man, and God and nature, and God and naturalization, could be easy to explain. This is through deformation. I believe that there is a force higher than God, that created God. Yet, this does not make a good theory. I am a Christian, who knows, that these things, are about the world, and about how the world, is about the nature, of man, and God, and about how the deformation, of earth, is about the philosophy, of how the earth started, with the big bang too? No. That one is misleading. The fact, that evolution is misleading, is the same argument that the Bible is not misleading. This is about the ways that the Bible compares itself, to the human aspect of Jesus Christ, even though He was God. He is about the ways that the world is about the ways that man is about God. This is about God, that is about God, that is about God. He is the one who knows things, and makes things happen. He is in the universe, so He is about the things, that He is about for a sure fire reason. He is the maker of heaven and earth. This is about how God, is about natural things, that is about how He is about service, that is about the salvation of mankind. He is about the creation that is about the ways that man is about God. God is about the man, who is about the man who is about the man, whom is surely

about man. He is the God, who knows about the God, who knows about the man, who knows about the God, who knows about the man. God is man. He is the collective, entity of man, that is about the collective entity, of man, that is about the God, who is about man. This is about the way the man is about man, who is about man, who is about man, that is about man. He is about God and man, who is about man, whom is about the man, who is always about man. The man, who is about the man, is about the elements that He created too. He is the one who fashioned earth after the elements, that were about man, and God's creation. Yet, the figure of God, in the beginning, is about creating the world for Himself. He was the one who made man and woman for himself. He is the one who knows about the way creation is about the ways that creation is about the elements too. This is because this is a scientific theory, on God. This explains the origins of the universe, and how we are part of this. This is a explanation, that is about the theory, that is about the God, who created us, all, in His Image. When, I went to church, in the past, when I was in High School, I believed in evolution. This is the theory, that I believed in. This is because it was taught to me. I did not ever explain how evolution came around, but I did see how it worked. This is through the theory that was not true. Yet, I still was taught this theory that did not make sense, to anything, in my life, except contests. These are usually about the surviving of the strongest, or more better, that is the opposite of Christ. This is about Christ, and how He was in the beginning. He was the one who knows about the collective, intelligence, of the world, and is about how He is about the collective, intelligence, of the world. He is about the one who is collective and knowledgeable that is about the ways that the new and the old, have gone and passed,

away, with the issues of God, and about how God, is about the ways that God, is about the essence, that is about the ways, that Darwin was in Christian, book stores. He was a liar, sort of, or else he was very strange. This is about how this theory is not compared to evolution, but is about the ways, that the man, is specific. He is about the ways, that man is about the ways, that man is about the ways, that man is about the ways, that is about the ways, that is about the ways, that is about the ways, that man, is about God. These are the proof positive, results of the word, that is about Jesus Christ. He is about the ways that man, is about the man, that is about the way man, is about the God, that is about the way, God is about God, who is about God. He is about God, who is about man, who is about mankind. These are all collective as the image, that God has created the world as. This is about the ways that man is about the ways that man, is about the ways, that man is about the ways, that man is about the ways, that man is about the ways, that man is about the ways, that man is about the philosophy, that is science. These are the scientific ventures, of mankind. These are ventures, of mankind, that are about the ventures of mankind, that are about the philosophical, ventures, of the modern day man, and his opinion. He is about the way that man is about the way the man is about the way the man is about the way man is about the way the man is about the man. This is about the way man is about man that is about man that is about the man, who is about man, who is about man, who is about man, who is about man, who is about mankind, that is about the Son of Man, who is about the control, of the world, through the entities, o f mankind, and about how the entities of mankind, are about the results that we conclude, are about ours, and the conduction of research we do, which is about God, and Jesus. This is

about reading the Bible. The Bible is about the ways that the Bible, is about the word of God, and about how the Word of God, is about the ways, that are about God, that are about the Bible, that is about the Bible. The only entities that are real are God and Jesus. This is from extensive research my whole life. This is about where they are, in the life, and the where, that they go to and come from, in the earth. The earth, and the sky, and the air, and the wind, and the water, and the fire, are about Jesus Christ, also. This is about the explained, results, of a experiment, that I have created, to be mine. This is that the elements also, formed together, when God's voice spoke, and when He was already there, the elements, were also in fact, present. These were the elements, of God, that were created, when God was created. This is why this theory is sound. These are the results, of the word, that is about the adventures, that are about the adventures, that are about the adventures, that are about the world's creation, even if the church doesn't believe in this, but might be interested in this, through science only. These are all about the sciences, that are all about the physics, that are all about the mathematics, that are all about the philosophy, and the one word that makes sense to all of this, known as "naturalization". The fact, at opinion, and factual existence, is about the real and the fake. These are part of the factual existence, and not just fact. Jesus was fact, and the natural realm. This was of His life and how He was the earth. He was in the body of God, when the world was created, and engineered by the mind, and the intellect, and the research of God. He actually, did not research, the world's beginning, but did expect, us to do this. This is why we were, created with a simple intellect. This is about the creation, of the world, with the glory of God, in many shapes and forms. He was considered the "ourselves" and

the "Spirit" that does not make sense, unless you research this, in whole, including essence, and substance, that is about the ways that God, is about man, that is about Jesus Christ, that is about the ways, that Jesus Christ, and God, are about each other. The Word was what became the flesh of man, which was in the beginning according to theology that studies this whole entire matter. This is about how lies, also form things, that are not Godly. This is why I think I will not venture this opportunity. The opinion, of God, is that He created the whole entire, world, that was in the beginning, anyway, that was only for Him, and Adam and Eve, which is fact. This is through the beginning, of the world, that is another reality, that we do not understand, but comprehend. This is about the way the comprehension of the world, is about the issues of faith, and how belief, and faith, do result in the beginning, of matter, and material, ways, that are about the philosophy that is about the man, and God, and man, and God, and about how the man and God, became with the world. This is about the becoming, of the world. This is through the "spark of the imagination" rather than the "spark of a bang". The meaning of the "big bang" is about the symbolic meaning of how the universe was created. This was all real. This is just if you choose to believe this, in matter, and composition, of what is around you, and what you believe in, which is about the components of the universe, in your opinion. Naturalization explains that the big bang was wrong, in exact terms, of naturalization, and how and why this is totally and completely wrong, perhaps, and with certainty, of misunderstanding, according to me. Darwin was misunderstood. He thought that we came from apes. The big bang would not have a composition of the universe, if it was an explosion. There was the bomb, that was dropped, on Hiroshima. This explains nothing.

The way the big bang works is that the world, according to me, is about the evolution of this. Perhaps, Darwin was right, and we all came from a nuclear explosion. According to quantum physics, the bomb of Hiroshima, was wrong, and deadly. This is how the big bang, was, relative to a piece, of bomb, that was the creation of the universe. This is about the way the Hiroshima, bomb, was about the constant and correct analysis done by scientists. If we study the big bang, relative to this, the big bang would be wrong. This is according, to naturalization, that is a safe theory, with friends and enemies. This is about how the abstract, is real, and the real, could not be an explosion, but a real bomb. This is what explains the logic, behind the big bang. This is that the world, could not be created with life, if a bang was responsible, for the world's creation, especially if we knew about it. The theories of science, do not follow a book, that is about creation. All it studies, is the elements, that are around and involved in, the essence, of substance, and formation, of thinking that is about the big bang and how its structure is impossible. This is with the naturalization of the elements, and how they are the formation of the universe, through God's voice, that spoke and created it. It is about nature, and man, and God, that explains the formation of the universe through naturalization. These are naturalization, and about how God, is a part, of this. He is part of the part of God. This is about the ways, that naturalization, is about, that, which is about the bomb that is about the logic. This is about the essence, and substance, of modern man, that could not have come from a explosion. This is of and about the way, the composition of the universe, is about, also, the naturalization, that is about the elements, that are how we formed, in the ways that we see, and hear, and how we survive. This is wit h the elements, that God cre-

ated the universe with. These are with the world, that was created by God. Yet, the big bang is caused by a spark in the beginning. So, this could not have been around. This is because of the truth in naturalization that is in the Book of Life, that explains that all are saved, who believe in Jesus Christ.√ This is about Christ, that is about Christ, that is about Christ, that is about Christ, that is about boredom with Christianity. This is caused by incidents of the big bang. This could have been a mistake, in my eyes. This could have been where evolution came from. The elements, of the composition, of matter, could explain, why Darwin, believed in evolution, the way, that we all evolved from an explosion. This is about the ways, that science, and math, are prevalent, and is about, the ways, that science, and math, are about the philosophy, of the essence. This is about the essence, and substances of math. The math that explains the world, is also through itself only. This is about how we cannot justify, the science and physics, behind the way the world was a voice. This was from, "Ourselves" which is my point, sort of, in the way, that we see the world. This is through Jesus. Jesus Christ, also created the world, through the voice, by being God, His whole life. This is about how the world, is about the creation, of the world, that is about creation, that is about the creation, that is about the creation. In terms of deformation, the earth formed, with the God, with the naturalization, with the essence, and with the substance, of man, because of Jesus Christ. Yet, if we are in Christ, we believe, that He created the universe, and not us. This is only if we are a lunatic, do we believe, that God created it through us. Naturalization, is an explanation, that God formed the earth. And, this is about the explanation, of how God formed the earth, that is of the world, theory of thinking, that can understand this, known as the

word, "naturalization". If this is explained fully, then the world, can be made sense of. This is merely a science attempt, with an art form, to understand the earth, and how it is formed, from God. This is also people. This is the kind of logic behind the human centered, devices, of the approaching apparatus, of the humanistic thinking, behind the creation of, a God-centered anthropomorphic, creation. Our God, is part human. He does not take on shapes and forms, of nature and the way, the world, is shaped, because of God. He is the absolute God, behind all Gods, and man, and men, and even Jesus Christ. Jesus Christ, is the absolute God behind all that happens on earth, and in heaven. These are the holinesses, that are in the constructions, that are in the conclusion, that is about logic, and about how, the logic, of earth, that is about water, that is about fire, that is about air, that is about wind, that is about sky, that is about the absolute methodology, that was about Jesus Christ, in the beginning. This is about the ways, that naturalization, is about the essences, that is about the substances, that are about the conclusion, that is about the ways, that are about the conclusions, that are about the naturalizations, that are behind the relevancy, that is about the way, the earth, formed, that is about common knowledge, that is about earth.

The Elements of Earth, and How the Earth Was Formed with God

The earth, according to the elements, that are about the intelligences, which are about the explanation, is about purely science. This is about the ways, that naturalization, is about the essences, that are about the substances, that are about the reasoning, that is about the compromise, that is about the earth, that is about the form, that is about the formation, that is about the conclusion. This is about

the obvious. This is about the coming conclusion of God, only. I do protect my faith with this book, that is not challenged, and is not debated, and is not compromised. But, it is purely a theory of science, that explains the way I think. The earth, is about the obvious, and necessary explanation, because of science, and about how the science, that is about the research, that is about the creation. This is about the way, that is about the ways, that science, and nature, is about the new school of philosophy. This is about the ways that is about the way, that is about the way, that is about the way, that is about the way, that is about the way, that is about the way, that is about the way, that is about the way, that is about the way. This is about the way that earth, was formed, and how Jesus Christ, was this formation. The "Word of God" was in the beginning. And, this is the way, that the world, is about the way, the world, is about the way the world, is about God, and nature, and man, and the environment. These are about the issues that are about the sense and the senses that is about the way, that senses, are about the conclusions, that are about the God, that are about the essences, that are about the substances, that are about our learning. These are the elements, that are about the substances, that control our world, through the human and nature and God and man and mankind and naturalization, and environment. This is not just the Gospel, but the written Word of God. He is the one who "spoke" and it was created. These are about the essences, and the ways, that human and God are about the essences and formations, that are about God and man. This is about God and nature and man and mankind and naturalizations, and the environment, that is always about our own knowledge, but with naturalization, in every sense, and in every way, that mankind can think. This is about the ways that mankind,

and man, and God, and nature, and environment, and naturalization itself, that I have not demonstrated, but are also, in the perspective, of man, are about the philosophy, and natures of man, which are in the conclusion. These are of the ways, that man have changed, and adapted, to a new way of thinking, that is about the intellect, and is about the ways, that the intellect, are about man, and about how man, is about the concise, and content definition of mine. This is not anti science or anti God, but a theory of the way the world was created, through the aspect of God. God is the naturalization, that is about the essence, and formation, that is about the conclusions, that we still draw and always draw. These are the ones about ourselves. These are for the math, that is about the science, that is about the logic, that is about the sense, of creation, that is about the concise, and intellectual ways, that man is about knowledge, and eventually is about the creation of the world, that is through the nature of man, and about the man, that is about the man, that is about the man. He is about the man, who is about the concise, and conclusive, orientation, of the earth. We cannot explain the origins of the universe, through, science, and the Bible, alone, but we can create a theory that is almost the same. This is naturalization, and about how the nature of man and God are about the conclusion, that is about the ways, that the conclusion, is about the essence, that is about the God, that is about the essence, that is about the fundamentals, that are about the conclusions of our life. These are the factual existence of naturalization, and how this is not about itself. This is about God. The ways that the structure of nature and man, are about naturalization, is about how the structure of man, is explained. This is always factual, and obvious, if explained the way that is right for man to see, and become with. This is the definition of

sane. These are about the conclusion, that is about the ways, that the conclusion, that is about the ways, that the conclusion, that is about the ways, that are about the ways, that are about the ways, that are about the conclusion, that are about man, and mankind, and God, and nature, and environment, and naturalization, and the best and the most obvious ways, that we can become. This is with God. This is about the ways, that God, is about the ways, that God, is about the ways, that God is about the ways that God, is about, the ways, that naturalization, is about the conclusion, that is about the way, that the nature of man, and God, is about the ways, that naturalization, is about the kind of thing, that is about the essence, of the formation, of God, and man, and the tales, of the creation of God and man that are about God. This is about the way that the God, of the universe, is about the ways, that the God, is about the way, that followers are about God, that are the substance of God, within the essence, of the surrounding Word of God, that was born in the beginning, of time, and the beginning of circumstance. These were the followers of God. He is Jesus Christ. He is the one who controlled the world, through His followers, that controlled themselves, through Jesus Christ, who was the one with faith, that they shared and followed, to become, "like Christ". They would, control, others, through Jesus Christ, to write in His Books, known as the Bible verses. Or, they were kings from the past, creating God, for the civilizations. These are about the ways, that justice, is about the way, that capabilities, are about the essences, of the substances, of the earth, that is about the elements. After Jesus came, He changed the elements, to His body. These are about the elements, and the way, they are about the controlled reality, of our world, unless you go Christian. Then, you are part of the creation of the world. The-

se are a part, of the world, that were created with this. These elements, are about the ways, that people, are about the special abilities, that are about, the control, of the world, that is about, the control, of the world, that controls us. This is because of the formation of the universe, where, capabilities, and elements, go together, and make sense, on a higher level, than before, when there was just people, who were like God, but not at all. In their, opinion, they were better than God, when He was betrayed. This was why He was betrayed. He was the perfect God, for everyone, to have personally, and socially, and basically, the best in the world, with followers. He was born of a virgin, so he was basically, from heaven, and the elements that He was born with, were gifts from the three kings, known as frankincense, and Mur, gold, that was the reason, why they, went to Him, and gave Him, these gifts. These were because of the world, that knew that He was "king of kings". Also, he was born in a manger. This is about the ways, that naturalization, is about, the elements, and the ways that the world, is about the conviction, that is about the essences, that is about the qualities, that is about the elements, of jewels, and how these are for a king, and the elements of the earth, are for anyone. These are the elements, of ours, for a king. These were the elements, that made sense, to these kings, to give Jesus Christ. So, they were like His only, His whole life, meaning that they were totally His, only. This is about a gift for the King, of creation, that was about the survival, of a God. This is about God, that is about the essence, that is about the substance, that is about the research, and how the Kings, just found Jesus, and went to Him. The elements of the world, somehow, control the world. This is about the elements, and of the elements, that are like the elements, that are about the gold, and frankincense, and Muir. They

are about the ways, that are about the ways, that are of and about, the ways, that people are of, and about, the essences, that are of and about the substances, that are and of, that which are about the elements, of everything, that is of and about the collection, that is of and about, money, that is of and about, the elements. These are what control, the world. This is through health. These are about the ways, that are about the way, that is and are about, the God, of the universe, and how He created, the elements, until Jesus came. He was for the issues, of justice, and compromise, and about, the elements, and how they were His, and His body, to share with others, instead of being selfish, and controlling them, and everybody, because the elements, are how the disciples, also had faith. They are what they believed in Jesus through. The basic, man, in the Garden of Eden, was created with a brain, a body, and a Garden. This is how he saw it. This was all his, and God's. This is until the Serpent, tested Adam, through the elements, known as the knowledge, of good and evil. This was in the shape, of a apple, which had knowledge of good and evil, which was his and hers, after they fell. They are about, the essences, that are about the substances, that were about a tree. The elements, were, sky, water, fire, earth, air, and wind, and they were in the knowledge. This was of good and evil, They were, about, the ways, that mankind, and man, and nature, and God, and naturalization, and environment, were about the ways, that we all fell, with knowledge. This is because of the ways, that God, created, the world, that is about the ways, that man, and nature, see this, only through naturalization, and its promises, and compromises, and abilities, and strengths, and the control, of the world, through these elements, that are of and about, the essences, that were about the qualities, that were. These were, always, about the way, the

nature, of man, was about the nature, of God. These are about the specifications, of man, like a resume, for a job, that he was created with, and about, and for. These were, about, the essences, and qualities, that were about, the conflict, and content, that were, and are. These are about naturalization, and about how, naturalization, is about the earth. This is how the earth was made, and how the earth was started, since the beginning. This is no matter what you think, or know, that is about the ways, that knowledge, is about the essence, and substances, that are about the quality, that is about the ways, that is about the ways, that control and essence, was about the substances, that were in the creation of the world. This is through a voice. Also, this voice, also relied, on the elements, that were created, by forming them this way. These were the creation. These were the elements of the creation, and how the elements, were about the ways, that are about the control, of the elements, and how He became with them. This is not superior to Him, but equal with His power, perhaps, in the explanation of naturalization, only, but God, is about the conflict and content, of the substance. This is about the real God. He is in the real Bible, and He is in the real, authority, of the world. These are about the ways that the world is about the ways that are about the God, of the world, and how He is about the conflict, of the God, and how the God was conflicted with God, before the creation of the world became. He is about the conflict, of these resources, too, which is when He created the world, also. He is about the ways, that the world, is about the ways, the world, is about the ways, that when the world, was created, became a world, for all. This is why He likes people. This is due to how He formed the world, and said it was "good". And, that is all He said, that it was, also because of the elements. These were in the

naturalization, equation, that helped God, and man, to create Himself, even, through the Word. We are Jesus' body, that we eat at church. We also drink His blood, or we would not be around. The opposite of these elements, are the real elements, that are about the elements, that were created, in the beginning by God. In comparison, the earth, and its elements, became Christ's, because of the body, of His, and the elements, He knew also, in the context, of the creation of the world, through a voice. So, He became God's Word. This was in the beginning, when the earth, was created by Him, also. He is God, also, and has performed many miracles, and operations, through these elements. There is holy water, (water) and raising from the dead (sky), and the breath of God, (wind) and His body, (earth) and healing disease (fire), and creating God (air). These elements, for Jesus, are the creation of the world. This is through naturalization. This is through the elements, that are about the creation of the world only. This is mainly what God, is known for. This is about the earth. If he created the earth, then he created His body. This was what was involved in the creation of the world, through the elements, that are the most influential intelligent items, ever compared to God's creation. This is the Holy Bible. This is the "Word of God" that was in the beginning. And, if this goes away, since Jesus Christ, came, there is a war. This is the way of war, which is about the client, and the student, coming together, to form one union. There are many, much awaited, games, that are played, and play each other, and this one is strategy. This is the way, that the games, of war, are played, and the way, the wars, are played, are about the client, and the student, participating, together, to end creation. This is about God's creation. This is a beautiful creation, that is about, the games, that we play, and how we play them. They are

about, the games, we play, and about how we conduct our research, and play them. This is about lies, and storytelling, that is about the honor, and glory, of the prideful, games, we play, and about, how these games are played, that are played this way. This is about the way, the games, are played, that are played this way. This is about the games, we play. This is about, how, we play them, and about how we are associated this way. This is about the way, games are played, and about how we play them, is up to us. This is about our games, and how we play them, is up to all of us, and each of us individually, that are shared and prospered, this way. This is in accordance, with the Bible. This is about each, game we play, and about how we play it, that is about it, and about the way, it is played. There are about a whole, many, games we play, and there are about a whole, many ways, to play the game. This is about the game, and about the way we play them. This is about the way God, plays games, and about the way He has envisioned the world. This is all for Him. This is about how God, plays games, with Himself, and how He wins. This is always, in the body of Christ, and how He is better than all of us, so in conclusion, why do we even play games. This is not in the God, that created our world. And, if God, plays games, it is us who are playing them. This is about the research, conducted, all through the world. This is about the game, we play, and about how we play, them, that is about how the world, is satisfied, and conducted, that is a part, of the research, that is conducted, in the ways that games are played, and how these games are played, are against, one another, and how these games, are played, is usually with intelligence. This is about games, that are about games, that are about the way, the games, that are played, are usually, played, with honor, glory, pride, and the product, of killing, and raping,

and stealing, and cheating, and the likes of the ways, that are about the games, we play, are about the conduction, of research, that are about the God, that researched this also, through man, but did not do this at all. This makes sense to the humanitarian aspect of the God of the universe. But, there is none of this. There is a God, that is about the God, that is about the game, that is about God, that is about the God, that is about the God, that is about the God, that is about the God, that is about the God, that is about the God, that is about the God, that is about the God. This is about the ways, that God, is about the way the God, is about the way, the God, is about the way, the God, is about the way, the God, is about the way, God, is about the way, God, is about the way, that God, is about the way, God, is about the way, that God is about the way, that God is about the way, that God is about the way God is about God. The way God, is about the way, that is about the way God, is about the way God, is about the way God, is about the way God, is about the way, God is about the way God, is about the way God, is about the way God, is about the way God, is about the way God, is about the way God, is about the way God, is about the way God is about the way, that God is about the way, God is about the way God is about the way God is about God. This is about the way, God, is about the way God is about the way God is about the ways that God are about the way that God are about the way that God is about the way God is about the way God is about the way God is about the way God is about the way God is about the way, that God is about the way that God is about the way that God is about the way that God is about the way that God is about the way that God, is about the way that God is about the way that God is about the way that God is about the way that God is about the God is about the way that God is about the way that

Riley's Natural Naturalization

God is about the way that God is about the way that God is about the way that God is about the way God is about God. He is about the divine province, that is about the concise, and bright ways, that people, are about teaching, and how people are about the learning, of possibility, and how the teaching, is about the lesson, that is about the ways, that are about the essence, that are about the essences, that are about the brain, that are about the conclusion, that are about the thoughts, that are about the conclusions, that are about the research, that are about the development, that are about wars, and how we learn from them. This is about how we teach each other to fight. These are the fights, which are about the teachings that are about the concise and right teachings and ways that are about the conduction, of fighting, and how we all learn, from this. This is about the ways, that are about the right and proper, ways that we learn, how to teach, that is about the ways, that are about the ways, that is and are about the conclusions, that are about the teachings, that are about the lessons, that we have learned, and are about the reasons, that are about the teachings, that are about the approach, that are about the ways that the lessons that we learn are artificial, and the lessons that we are known, to have learned, are a product, of civilization, and are about how, the teachings, of the lessons, are about the approach, and concern, of a body of Christ, that is about the lessons, that we have learned, and about how we have learned, that is about the lessons, that we have learned, that are about the corrupt, and insane ways, that the teachings, of the lessons, are not about God at all, but about the lessons, of the corruption of the system. These are the teachings, that are about the teachings, that are about the way, that are about the lessons, that are about the ways, that the lessons, that are about the ways, that the opinions, are

about the ways that the lessons, are learned, and are absolutely distanced between two people, and the approach of the two people, that are about the body of corruption, that is about the dichotomy of good and evil, and about how these ways are about the corrupt, and insane ways, that we approach democracy, and the communism, and the ways that the government, follows, the real and fake, ways, of salvation, and how we are a product of our humble civilization. These is about the ways that the product, is about the industry, that is about the product, that is about the research, that is about the concise, and difficult, direction, in the ways the way follows the search, and the way that the bring forth of direction is toward the goals. These are about the ways that the goals are toward the direction, and the ways we follow, are toward the direction of the center, and about how the center of the ways, are about the approach of the circle, and the ways that we are governed, by this approach, is about the circle that is about the direction we follow, that is about the governing body, that approaches, the circle. These is about the ways that the circle and the direction of the circle follows, the same direction, that is about the approach, and is about the angle, that is about the common denominator, that is about the ways the approach, will consist of and follow, the direction, of the necessity, and the following of the circle toward the direction of the circle, that is about the following, of the earth, that is hemispherical and circular, in the ways that the directions, of the stars follow, and how we get gravity, is through the approach from science. These are the ways, that we follow, into the realm, of the following, into the realm of education, into the realm, of government, into the realm of church, into the realm of the difficult measures that are into the body of Christ, named the body of Christ, that is the body of Christ, that is

the body of Christ, that is the body of Christ, that is the body of Christ, that is the body of Christ, that is the body of Christ, that is the body of Christ, that is the body of Christ, that is the body of Christ, that is the body of Christ, that is the body, of Christ, that is the body of Christ, that is the body of Christ, that is the body of Christ, that is the body. This is of the church. This is also of the government. This is also of the school, that is about the education, that is about the school. This is about the holiness, of the body of Christ, that is about the concise and concluded results, of the game. This is about how we are in church, and how we are in school, and how we are in government. This is through the body, of Christ, that man has blessed himself, with, that is true with the church, that is truth with the government, that is truth with the school system. This is about the conclusion that everyone is about the body. This is of God. What else would form a body? This is the absolute truth. This is how the absolute truth manifests itself in society, and through the games we play. This is not God, but our body. This is eventually Jesus' judgment. This is directed at us, and not Him. These ways, that are of and about, the essence, that is of and about the substance, that is of and about the conclusion, are of and about the ways that man is about the context, that is about the man, that is about the ways that man is about the conclusion, of the body, of the government, that God also created. This is through naturalization, eventually, or what we have called the modern day period, of dramatic change, that has always been changing, into more modern. These are the ways, that the conclusion, is about the threshold, that are about the cheating, stealing, and lying, that are about the concerning issues, that our modern day, has always faced, that is through the system, of change, and the dynamics, of the interwoven structure, of the liv-

ing and the dead. This is about how Christ, has judged, the modern day. These systematic, of the methods, that are about the ways, that are about the concern, that are about the product, that are about the concern, that are about the disillusionment, that are about the science, that is about the math, that is about the concern, for others, that are about others, that are about the for and against, that are about the creation, of the product, that are about the research, that are about the approach, that are about the concerning natures, of man and beast, are about the concerning pressured issues, that are against and for, man's challenges. This is always about the man's nature and his concern. This is for the issue of God, and how we deal, with it. This is in church, and is about the ways that government, is for only one, and only one of these, is about the ways that the system, is pressured and is about the ways, that the society, is for the good, and the social issues that guide our country. These are about the God filled dilemma. This is about the complex, of the context, that is about the consistency, that is about the consent, that is about our modern day, trilogy, of God, Jesus, and the Holy Spirit. These are about, the issues, that are about the realm, that is about the issues, that are from and with God, that are about the God, that are about the God, that are about the God, that are about the God, that are about the God, that are about the Jesus Christ, that are about the Jesus Christ, that are about the Jesus Christ, that are about the Jesus Christ, that are about the Jesus Christ, that are about the Jesus Christ, that is about the Jesus Christ, that is about the Jesus Christ, that is about the Jesus Christ, that is about the Jesus Christ, that is about the Jesus Christ, that is about the way Jesus Christ, is about the Jesus Christ, that He is, and He was, and He will be. This is about the concern, that is about the creation, that is about

the contextualization, that is about the creationism, that is about the concentrating issues, that are about the creation, that is about the disillusionment, in which seems to be about, the control, and the policies, that control us, that are about the development of the idea, that is about the content, and the way we know this, which is about the bomb, which is about the explosion, that is always about the trust, and issue, involving the issues and science, of the research and development, that are about the philosophy, and the conclusion that are about the context and content that is about the conflict and interest that is about the emotional pleasure, that are about Jesus Christ, that are about the context and content, of the approach and proliferation, of the world, and how about the world, that is about the world. These are the issues of Jesus Christ, and the guiding light, of saving grace, and how we become, what we want to be ending with. This is up to us. Yet, we are a body our whole lives. This is directly, the body of Christ. This is the about the way, that the issues are, found, and about how we accept, our issues, with the body of Christ. The details are irrelevant. These are the approaches, that are about the approaches, that are about. This is the body. This is the body of Christ, and the elements of desire, that are prevalent in the world. This is the design of the world, that is about the way, in which, even the perpetual axis's are bound to be part of, so there is a structure to the church, perpetually bound by desire. This is about the desire, perpetually bound by human nature. This is about the ways, that the body, of the Christ, is bound by perpetually desired, things, which is bound by persuasion. These are about the essence, that is about the desire, that is about the desire, that is about the ways, that desire, that is about the influence, that is about the desire, that is about the desire, that is about the axis, that is about

the circumference, that is about the equation, of relativity, and how the earth is bound by perpetual motion, because of this. This is the death toll of the Garden of Eden's fall. Yet, if we were timeless in the Garden of Eden, then we would have perpetual motion, that is bound to the axis, of the grounds of persuasion, that are in the Garden of Eden, which explains why there is a Satan, and there is a predominant force, that is about the equation. This is about the ways, that man, and nature, and God, can have been related. This is due to the elements, of man's desire. These are about the ways, that are about the ways, that are about the ways, that are about the ways, that are about the ways, that are about the ways, that are about the ways, that are about the ways that are about the ways, that are about the ways, that are about the ways, that are about the ways, that are about the way. This is about the way, the dynamic, is about the circumference, that is about the direction, that is about the way that is about the dynamics, that are about the ways, that are about the ways, that are about the ways, that are about the ways, that are about the ways, that are about the ways, that are about the ways, that are about the ways, that are about in which are about the ways. These are about the philosophical ways, that are about the ways, that are about the ways, that are about the ways that are about the ways that is about the way that is about the way that is about the way that is about the way that is about the way that is about the way that is about the way that is about the way that is about the way that is about the way that is about the philosophical ways that are about math and science and God, and the ways that movements are about the way that are about the way that are about the way that are about the way that is about the way. These are about the coordinates on an axis. These are the elements and how

they are conducted and researched and approached through the axis. These are the approaches that are about the way the conduction of the experience is about the direction that is about the direction that is about the dynamics that are about the approach that are about the dynamics that are about the essence, that is about the way. This is about the way that the world is about the way that are about the way that are about the way that is about the way the world is about the dynamics that are about the way that the dynamics are about the way that the dynamics are. These are about the consequences. These are of little of no demeanor that is about the control, of the environment, that are about the sense, that are about the sense, that are about the sense, that are about the sense, that are about the sense, that are about the sense, that are about the sense, that are about the sense, that are about the sense, that are about the sense, that are about the sense, that are about the sense, that are about the sense, that are about the sense, that are about the sense, that are about the sense, that are about the dynamics, that are about the sense, that are about the sense, and the ways that the sense, is about the ways that the sense, that are about the sense, that are about the sense, that are about the sense, that are about the sense, that is about the senses that are about the senses that are about the senses. These are about the sense, and the ways of the sense, that are about the ways that the sense, is about the sense, that is about the sense, that are about the sense, that are about the sense that are about the sense. This is about the miracles, and wonders, and signs, that are part of everyday life. The Bible, is a tool for the senses, of mankind, that are about the ways that man, and mankind, have and see the senses, that are about the sense. This is about Jesus Christ. The other senses, are the elements, that are about

the ways, that the senses, are about the accumulation, of misdeed, and the ways and paths to the creator of the world. This is through Satan, or the fallen angel, from heaven, and his accumulation of senses, also. These also include the senses. These are about the ways, that the senses, are about the approached fall, of man.

Satan, has power over wind. This is natural. He is the prince of air. He is the promoter of fire, known as hell. He is the one who walks the earth. He is the false God, or the air prowler, who crawls through the air, in disguise. He is the victim of water. He is the creator of the fallen idol, which is all about stars, and astrology, and reading falsely into the universe, in the sky. And, he is the victim of the tragedy, that is known to man, as the stage. He is the world leader, who frolics along the open edge, of danger, that is about the way, that he is about the way that he is about the way that he is about the way that he is about the way that he is about the way that he is about the essence, and substance, of man, that is about mankind, that is about the God, of the earth, that is about the God, and what the, God, does, is about the way, in which He knows, about the direction, that we follow. He is about the ways, that the God, of the earth, is about the ways that the God, of the wind, is about the way, that the God, of the earth, does not deal with the elements, but instead created them, to be His, before the world was created, and just knew how to speak, unto the earth, to create it. Yet, He did not create it, but his voice did. This is about lies and how they do not believe this. This is about the ways, that the world, but one person, Jesus Christ, was created. This is because Jesus Christ, is not a skeptic. He is a creation, of the earth, and what the earth, is about, and is because of is because of the future. He is like this, in that He is about the way,

the world, is willing, and able, to create the world. He is the skeptic. He is the one, who is about and of the world, that is about and for, the world that is about and of, the skeptic, and the way that the skeptic, is about the way, that Jesus Christ, is about the way, that Jesus Christ, is about and from, the open book, that is about and from the world. This is about the way that the world, is about and from, the earth, that is about and from the way the world that is about and from, the world, that is about and from the earth, that is about and from, the earth, and is about and from, the earth, that is about and from, the earth, that is about and from, the earth, that is about and from, the earth, that is about and from, the earth, that is about and from the earth that is about and from the earth, that is about and from, the earth, that is about and from the earth, that is about the way, the earth, is about and from, the earth. This is about the way the world, does things, that is about the things, that are about these things that are about the things that are about the way the world does things, that is about the aspects, of nature and man, to do things, that are about the aspects, of nature and man, to do mankind, things, that are about mankind, that are about mankind. These are about mankind, that are about the ways, that mankind, from the perspective, of mankind, is about capable and willing things. These are about the constituents of man, and mankind, that are about, mankind, that are about mankind, that are about mankind, that are about the mankind, that are about mankind. These are about mankind, that are about, and from, the mankind, that is about the way mankind, behaves, that is about the way mankind, behaves, that is about the way mankind, and his philosophy, is about the way mankind, is about mankind, that is about mankind, that is about mankind, that is about mankind. These are

about philosophy. This is for mankind, that is about mankind, that is about mankind, that is about mankind, that are about mankind, that is about mankind, that is about the word-specific tales of mankind, that are in the book. These are about reactionary mankind. These are the correct words, that are about mankind, that are about the product of mankind. These are about mankind, that are about the way mankind, is about mankind, that are about mankind, that is about the extremities of mankind, and the ways mankind is about thought that is about the ways mankind is about thought that is about the way mankind is about thought that is clear and simple ways of mankind. This is naturalization. This is the clear thought that is beyond mankind. These are the beyond, mankind, that is of and about, mankind, that is of and about the natures of mankind. These are prevalent in the naturalization, theory. These are about mankind and are about the ways mankind are and are from, mankind, that is clearly about the philosophy, of mankind, and its generosity. This is about the prevalent appeal of mankind, to the real standard, of naturalization. The real generosities, of the standard of mankind, is about the clear level, of thinking, that is and are of and about the clear level of thinking. I am typing this so you can know where I am coming from. I am trying to communicate a clear level of thinking. This is so that you cannot judge me. This is about the level of thought, in naturalization. This is about the clear, level, of thinking, that is and are about the clear level of thinking, that is and are about the clear level of thought. This is about the clear level of thought, that is about the way the thought is about the clear level of thought, that is about the level of thought, that is about the level, of thought, that is about the clear level of understanding, that is about the way, that thought, is about the level of thought, that is

about the way the world, is about the way the world, that is about the way the world is about the dialect of reason, and about how, the reason, and response, is clearly about the communication. This is why books are pointless. These are about how they are read, and used. This is except for the world, that is about the clear and purely, responsible, ways of thought, that are about the tactic, of the world, that is about the way, the world, is about the way, the world, is about the ways, that man, and God, are about the clarification, and responsibility wise issues, that school poses. This is also, about the dialect of church. This is about the response, of government, that is about the clear and clever, thinking that is about the clear and clarifying, governmental response, that are about, the clear and present, thinking, which is about the clear, and the level, of thought and clarity, that are about the response, and clear, thinking level, that is about the psyche. This is about the way, the basic level, of understanding, is about the clear, and response, capable psyche, that is about the level of thought, that is about the clear level of thought, that is in the world. There is a universe that is about the clear, level of thought. This is called heaven. This is where the final outcome, of the chapter, is the about, and from, of the conclusion, that is and are of, the outcomes, that are of the final, conclusions, that are about, the book, that is of the Book of Life. This is where the body, of Christ, is about the concise, and outcome, ability, of the clear and response, thought, defies thought, and goes, to the ending. This is about the ways, that are about the ways, that are about the ways, that are about the ways, that are about the ways, that the world, dictates, and analogically, completes, the essence, and substance, of the ability, to choose between man and woman. This is about the dictation of circumstance. This is the final di-

viding line between knowledge. Can, God be man or woman. The answer is simple. God is made in the image, of man, if you accept this. The answer, to this riddle, is obvious. The answer is that God is not actually, man, but made man, in the image of Him. He made Eve, afterwards, from Adam's rib. The conclusion, of this context, is about the way, the man, and his counterpart, Eve, were created purely together, that is about the way the world is about the context, that is about the naturalization, that is about the context, that is about the universe. The Gods, who created the universe, spoke, and man and woman, became. This is about the way the world is about the way the world is about the way the world is about the way the world is about the way the world is about the way the world is above, and beyond, the reason, that there is a trial between, God and man. This is due to the fact in which that man and God, are about the conclusion, that man and God, are about the content that are conclusions, in which are made of factual existences. This is always my conclusion, when opinions are always what the real answer is made of. This is the whole of existence.√ This is always about the ways that man and woman always appeal to the docket of one end side or another shelf, per se, of the solar system. These are elements that are about and from the ape to the God. These are the theories: naturalization and evolution. The docket appeal is from a Creed song I heard once with perfect faith and existence that is about and from the existence that are from the earth to the solar system, in perfect communication. This is the idol of perfection. This is the lazy mind and the homework you have to do. This is about Creed, instead of perfection. The ape was the perfect example of the human, when it tried to eat, sleep, or do anything. They are about eating, sleeping, and grazing, when they are energetic. They are about the

come, and do anything approach to the future dynamics of man, and of mankind, and of the man, that is of and about, the real mankind, that can and will, and do and does anything. These are about the conclusions, of the man, that is about mankind that is about man that are of and which is of and about and of and about and for mankind. This is of mankind. These are the variables. They are the elements, of naturalization, and not the substances, that are of which are the progress, of the equation, that is of and is and are, about and for, the substance. This is of naturalization. This is about the perfect, and willingly, capable, issues of mankind, and is of which is about and capable: which is of mankind. This is about and from, and capable, of being, that which of is, and about, the dynamics of man. These are the elements. These are for the capable elements, that are of and is about, the elements, that are of and is about, the dynamics, of man. These are always mankind. The about and from, of and about mankind, are about the essences, of mankind. These are about, and of, mankind. These are the issues, of and for and about, man, that is of and are about, mankind, that are about the issues that are about mankind, and the issues that are about the ways of mankind, that is about the way, that is about the way that is about the way that natural ways of mankind that is about the ways of the ways of mankind, are about the ways that mankind, and nature, of mankind, that is about and of the nature that is about and of mankind, are about and of the philosophy, and nature of man, and mankind, and the clear and present mold that are off the block of real existence, is the mold of preservation.√ This is about the ways that man and mankind, are about the ways, that mankind, are about the ways, that mankind, is about the way man, and mankind, that is about and of and about the man, that is about the mankind, that is

about the man, and the mankind, that is about the ways that man, and mankind, are about the functioning, that is about the capabilities that are about the philosophy, that is about the philosophy, that are about the ways, that are about the ways that man, and woman, are about the creation of the world, and God is about the creation of man. The way I believe this to be true, is of the capable resistance that is about God, that is about the way God, is about the progress, of the nature, of God, that is about the ways, that nature, and man, are about the way, that nature and man, is about the philosophy, that are about man. This is about man, that is about man, that is about woman, that is about woman, that is about woman, that is about man, that is about man, that is about woman, that is about woman, that is about man. These are about the ways, that woman, are about the way, that man does things, that are about the things, that are about the things, that are about cultivation, that is about skills and abilities, that is about the clever and clear, structure that is about the God, and man, of mankind, and about how mankind, and God and man, is about the ways that man, and mankind, are about the God, of the world, that is about man. This is about the way man, is about the God, that is about the man, that is about the God, that is about the way man, is about the God, that is about the God, that is about the man, that is about God, that is about man, that is about God, that is about man, that is about God, that is about God, that is about man, that is about God, that is about the way God, that is about the God, that is about the God, that is about the way, God, is about the God, that is always had man, that is about man and are about God, that is about the way God, is about the man, that is about God. This is about the way, that God, is about man, that is about God. The ways that man and God are about man is about the ways man is

about God. This is the way I conclude this chapter. The chapter concludes with an ending remark.

"Without God I would be nowhere".

This is about how the essence, is about the condemnation that is about the essence that is about the conclusion. This is about the essence that is about the conclusion that is about the conclusion that is about the essence that is about the conclusion that is about the problem. This is about the way, that the conclusion, that is about the essence, is about the conclusion, that is responsible for the essence, that is in the conclusion. This is about the way that God, is about the earth, that is about the conclusion, that is from and about the conclusion, that is about the conclusion, that is about the real conclusion, that are about apes and monkeys. The fact, that monkeys, exist, is true. This is about the fact, that Darwin was a lunatic. The fact, that monkeys, do exist, is about the fact, that they are monkeys, and that they are monkeys. This is the fact, that they are monkeys, that are monkeys, that are about the monkeys, that are about the conclusion. This is that the world, isn't religious, that is about the results, that is about the progress, that is about the conclusion, that is about the progress, that is about the monkey, that is about the knowledge, that is about the coming conclusion, that is about the fact, that is about that we are all equal, that is about the belief, that is about the customary belief, that is about the strange superstition, that is about the trial of Darwin, that is about the belief, that is about that the monkey, that is about God, is about the belief in Jesus, that monkeys cannot exist. This is unless you believe that evolution is true. This is in the form of my deformation. To relate this to the theory of naturalization, would mean, and signify, that the monkey is about the ape that is about

the monkey that is about the ape that continues forever. The belief would be in Darwin and that he was on trial, when he was on the island, believing in things falsely. This would signify that the belief was a opportunity to know where we came from. This was from the dinosaur age. This was from the big bang. In deformation of my own, the Bible says that the monkey, was part of the ark of Noah, that is about the belief that is about the conclusion that is about the context, that is about the content, that is about the consent that was about the belief that was about the know, that is about the belief, that is about the belief, that is about the belief, that is about the lies, that is about the truth, that are about the lies, and cheating and stealing, that are about the knowing, that is about the knowledge, that is about the knowledge, that is about the knowing, that are about the philosophy, that is about the "love of wisdom" according to my community college professor, that was an ape. This is because I believed in evolution. This was until I wrote about philosophy. This is why the evolution of the world, is about philosophy. This is about the elements, that are about the elements, that are about the elements, that are about the elements, that are about the conclusions, that are about the ways that we by, trial and error, create and condemn our own worlds. These are from very strange beliefs. To believe in naturalization, wholly, would be to not understand the belief of God, and the satanic beliefs of Satan, and the ways, that man and woman, are about the ways that mankind, and womankind, are about the fall only. This is through everything in naturalization. This is about the ways, that naturalization, is about the man, in which is about the God that is about the God, which is about the man, which is about mankind. This is about the ways, that are about man, that are about God, that are about mankind that is about man

and God, and are about the ways, that man, and God, and man, and God, are about man, and God. These are about God and man. To believe in superstitious beliefs, would be the answer to Darwin, and would be the end of the world. This would be the democratic, conclusions, to all, of the whole entire world, and that are a part of the grand illusion, one of the aspects, at a time, and would conclusion demonstrate, the essence of substance, and the willingly able and capable issues, of mankind. The ways that are ready and able, are about the ways that capable and willing people demonstrate the life of the world to ending and beginning and together demonstrate the will and the right and privilege to survive, the ways that we all are about and from, with instinct, that the preserve, and fitting and rightful, ways that demonstrate, abilities is through the rightful and ready, and willingly capable ways of justice, that can and will survive, to the end. The way we interpret the ways that the capable and willing, and the ready, and willing, and capable, ways, of instinct and preservation and survival and fitting and willing ways, to the communication, are the ready and willing, ways, to preserve, and remain steadfast the willing and capable ways of the world. This is about the way the world is about the way the world is about the context and content of the grand façade that is also about the ways that nature and man and environment are the about the ways that God and mankind and naturalization and environment, can survive, especially with the lost cause. The nature of man is about the environment, that are about the ways, that man and mankind, are about the naturalization, that are about the ready, and willing ways, that nature and mankind, and nature, and mankind, that are about the naturalization, that are about the essence, and substance, of the God, and willing and capable ways, that are ready,

and willing, for anyone. These are the chairs of excellence. These are the chairs of the righteous. These are the chairs of the worthy. These are the chairs of the great. These are the chairs of the honorable. These are the chairs that are about the willing, and capable, ways that college, and education, and government, are about the chairs of the next big thing. One monkey, can pretend to know what he is doing, and another monkey can see what he is doing, and another, can act like he is doing all. These are the powerful words, that can form, and are capable and willing, to accept, the honorable, mention, that are and is about, the collegiate, and governmental, and church person. The next thing is to accept to bring forth a challenge. All people, can do all things through anyone, that is and can be, honorable, righteous, and excellent, and can and will be the great, and noble, person, that is and are, a part of the government, and the chair, and the school, and the chair, and the great, and the chair. Yet, anyone can be accepted, into the church, if they stand on the right chair. This is the exit to school. This is the stage, to the monkey, and to the ape. This is the right turn, to the monkey school, and the monkey chatter, and this is the made up answer, that is of the monkey person. This is about, the philosophy, that are about the, essences that are about, the substances, that are about the chatter, that are about the monkey, that are about the excellence. If there were snakes, instead of monkeys, the snakes would go around everywhere, and deceive everybody, and then put it down, as "truth". This answer, to everybody, is the answer, that is about the monkeys, that is about the monkeys, that is about the monkeys. This is about the answer, that is false. This is, about the answer, that is and are, of and about the answer, that is of and about the answer, that is of and about the answer, that is of and about the

answer. This is of, and about, the answer, that is of and about, the answer, that is of and about, the answer, that is of and about the answer, that is of and about the answer, that is of and about, the answer, that is of and about, the essences, that is of and about the evolution. This is of and about, the conclusion that is of and about the answer. This is of, and about, the dialectic. These are always, of and about, the real conclusion, that are of and about, the answer that is of and about, the answer, that is of and about, the answer. This is of and about the answer. This is about the answer, that is about the answer, that is also, about the dream, that is about the awakening. This is about the God, that is about the conference, that are about, the ways, that are about, the, ways, that are about the Christian, mottos, that are about the conference, that are about the gatherers, that is a lot about the confusion, that are about the ways that come. These are the ways that we can come, into the society that are about the lesson. This is about, the came, and came, and then come, and then come, into society. These are about the lessons, that are about the lesson. This is about what, teaching does, and about how society, does teaching, that are about society, that is also, about, the timing, than teaching that does this. This is through the mind, of the element, that is through the elements, that are about these elements. These are about the elements, that are, about, the essences, that is constructed of the with, that is about and of, the essences. These are about the essences, that are about the essences, that are about the conclusion, that is about the lesson. These are about, the ways, the method, is about the lesson, that method, is from, and about, in essence, that are good, and are good, that are good, that are about teaching. These are elements, of desire, that we fell from, and about, and within, and within again, and without, the

pride, that consists of desire, that we are fall from, that these are about, these teachings, that are about, the client, that are about the lesson, that are about the lessons, that are about the lessons, that are about the ways, that are about the lessons, that are about lessons, that is about the confusion, that are made. These are through the elements. These are from desire. This is about desire, that makes, the desire, that is of, and about desire, than others, that call my bluff, cannot know about, and of, that is about the way the desire, is about the lessons, that is about, the lessons, that is about the crucifixion, that are about the ways, that came from, the earth, that is of, earth, that are about the conflict, that are about the interest, that are about the lessons, that are about, the lesson, that are about the teaching. These are about, the learning, that are, of and about, the learning are and within, the teaching, that are about the lessons, that we teach, and learn. This are about the way, that we know, and about with, in and of, the essence of substance, is about, the lesson, that are learned, that is of the knowledge, that are about the lesson that is about the ways that are about the almost situation in the world, that is obvious, that are in the way, that the obvious are about lying. These are about the lesson that is about learning that is about the conflict, that are about the disorder, that is about the lesson, and are. These are the confliction, and opinion, of the word of God, that comes in and from, the lesson, that are about the lesson, in which is about, the lesson. We are all reached, in the goal, of education, and the way, that lessons, are taught and is about the learning, that are about the God, that is about the conflict, that are about the conflict, that are about the issue, that are. These are from the God. The God, is about the way, the learned see, and with this, become. This is a math and a science, form of becoming, that

are about, the ways, that nature, and man, and are about, the conflict, that is about the empty, and empty apartment, that is about space, and time. This does, the image, of God, that is about and of, the essence, that are about, the lesson, about the social habits, the image, that is have, to do, with the attitude, that is done with this issue, that are about the image, that are about the image, that are about God, that are about the image, that are about the image, that are about God, that is about the image, that is about the image, that is about the image, that is about the image, that is about, the images, that are about the images, that are about the images, that are about image, that are about image, that are. These are, about image, that is. These are about, the ways, that image, is about the image, that is about the images, that are about the of, and about, the image, that are about image, that are about the image, that is about the image. These images, that are about the images, that are about the images, that are about the images, that are about the image, that are about the image, that are about the image, that are about the images, that make up our image. The image, is the image as the image, that are about the image, that are about the image, that are about, the image, the image, and the images. The ways, that we make up our image, is about the image, that is from the image, that is from the image, that are about the image, that are from the essences, that are about the lessons, that are about the lessons, that are about the God, who is from, and about, the image, that is and are about, the image, that are about the image, that are about the image, that conclusion is also about, that is about the lesson, and the way, the lessons, are about the ways, that is image, and about image, is not about image. These are about the images. The lessons, that are learned, are with, and in, the lessons, that with, and in, the essences, are

about the crucifixion of Jesus, and how He became God's image, that lasts forever , and ever, and ever, and ever, and ever. This is about how the crucifixion, of school, and the mascot, was about the main number, of people, who is about the ways, that is about the crucifixion, that are about God, and God, about the lessons, that are about us, and the ways, that we have, essences, that are about, the main number of people, whom are about the lessons, that are learned, with and from, the lessons, that are about the main number, that is about the way, that the main number, is from and about the God, that is from God, that is from God, who is the Father, and the Son, and the Holy Spirit. That way, is the evil, and wrong, ways, in which we all, learn and know, about in the conclusion, of the lessons, that are about the main lessons that are with God, and Jesus Christ, and the God, and Jesus Christ, and the God, and the real, Jesus Christ, and the Son, with the Holy Spirit. These are, about, the lessons, that are learned, that are about, the lessons, that we are learned, with about, the lessons, that we know, about and learned, about with the way that, the number, of many is about the number of many, that is about the number. This is the sign of the six, six, six, and how many people there are involved in it. This is about the understanding, of the Revelations, that I kept, on looking, on and at, and the ways, that the main number, was about, the number, of false Christians, there are and how they are about the many. These are about the many numbers. These are all calculated to be a six, six, six, and nothing more. They are about, the lessons, that are about, the main number, that are about the lesson, that are about, the lesson, that are about the lesson. This is about the, Holy Spirit, and how the lessons, that the way that we learned, are about the lessons, that are about the triple six, there are way the lessons, that are about the

six, six, six, that are about the creation, that are and is about the lessons, are about the creation, and the ticket to destruction, in Revelations. The way that maker of earth Jesus Christ is about, the way are that is about the lesson are about is about the are about that is about the are about, the lesson, are the ways, the lesson, are about the lessons, the which way, is are, and are, is about, is whichever direction, which is in business, that is in the number, of the mark of the beast. The, number, is, six, six, six, that is about the main number, of people, who are called to be with the beast. The way to beat the beast, is to pray, and fast, and see the number, of the beast, that are about the main, people, that are about, the lesson, that are about the mark of the beast, and these ways, that are about people, are about the way, they fast, and hold secure, to the way, that the beast, was about the number, and how we all fasted, in the way, that the beast, would lose. One, that you lose, is one that you lose, that is about the one, that you lose, that is about the one, that you lose, that is about the one, that you lose, that is about, loss of salvation, that is about one that you lose, that is about one that you lose, that is about, one that you lose, that is about one that you lose, that is about one that you lose, that is about one that you lose, that is about one that you lose, that is about one that you lose, that is about an expression. This is about the way, the one that you lose, is about the one that you lose, that is about the one that you lose, that is about the one that you lose, that is about the one that you lose, that is about the one that you lost, eternally, that is about the one, that you lose, that is about the one that you lose, that is about the one that you lose, that is about the one, that you lose, that is about the one that you lose, that is about the one that you lose, that is about the one that you lose, that is about the one that you lose, that is about the one

Riley's Natural Naturalization

that you lose, that is about the one that you lose, that is about the one that you lose, that is about the one that you lose, that is about the one that you lose, that is about the one that you lose, that is about the one that you lose, that is about the one that you lose, that is about the one, that you lose salvation in. This is the beast. This is of and about, the real beast, that is of and about, the real, beast, that is of, and about the real beast, that is of, and about the real beast, that is of, and about the real beast, that is of, and about the real beast, that is of, and about the real beast, that is of, and about the real beast, that is of, and about the real beast, that is of, and about the real beast, that is of, and about the real beast, that is of, and about, the real beast, that is of, and about the real beast, that is of, and about the real beast, that is of, and about the real beast, that is of, and about the real beast, that is of, and about the real beast, that is of, and about the real beast, that is of, and about the real beast, that is of, and about the real beast, that is of, and about the real beast, that is of, and about the real beast, that is of, and about the real beast, that is of and about, the real six, six, six, that is of and about, the real mark. These is about eternal. The way, the real beast, is about the real, beast, is often seen, as hypocrisy, that is, also, about, the enemy, and about how he is Satan, and how we must protect, and serve our community, for the earth to like us. These are about the equations, that are about the real beast, that are, about, and of, the real beast, that is about, the real beast, that is and are, about the real beast, and are about political power, and the mark of the beast, which is of the six, six, six, that are about, the, lessons, that, we, learn, and are, about, the lessons, that, we, learn, that, are, about, the mark, of the six, six, six, and how the number, is about the beast, and his mark, that will count, to six, and then die. The

symbolic meaning behind the six, six, six, is about, the lesson, that we, learn, and are about, the lesson, that the lesson, that the lesson, that the lesson, that the lesson, that the lesson, that is about the lessons, that are about the lessons, that are about, symbolic. This is, and are, the about, the beast, that are about, the real beast. This, is about, the real beast, and about, how the count, is about the from and about, of the six, six, six. This is about the way, that the beast, that the beast, that the beast, that the beast, that the beast, that the beast, that the beast, that the number of the six, six, six, is about the numbers, of the much, and many, that are about the followers, that are about, and are of, and in, the numbers, that are about the explosion, into the earth, that the beast, will come with, and devour. This is the big, and the small. The ways, that the explosion, of the beast, will happen, is through overnight, because of the big beast, that lives under our head. This is about, the idol worship, of a snake ring, and about how, the idol worship, is about, the man, behind the circus, and how about he, is about the ways, that are about the ways, that are about the ways, that are about the ways. These is about the mark of the beast, and how I am not a part of it. This is the conclusion of this book. This is about the way, that the beast, was about the way, that the beast, was about the way, the beast, was about the way, the beast, was about the way, the beast was about the way, the beast was about the way, the beast, was about the way, was about the way, the beast, was about the way, he lost. The way, that the beast, is about the way, the beast, was about the way, the reality of the beast, was about the way, the beast, that was about the way, that the beast, lost, is about the way, the police, are about the way, the police, are about the way, the beast lost. There are issues about this, in the media. These are actors, that can

think, that they are better, than the beast, when they are corrupt, and insane. The real place, where the image, of the beast, was created, was with the beast, that is about the way, the beast, was about the way, Hollywood, is corrupt. This is an, analogy, for the beast, that is about the way, the beast, that was about the way, the beast, was about the way, the beast, was about the way, the beast, was about the way, the beast, lost. This was, to, the, meaning of the symbolism, that was behind the beast, and the song. This was, about symbolism, that is about, the symbolism, that is about the symbolism, that is about the symbolism, that is about the real image of the beast. If you worship, him, you will contain a mark of the six, six, six, and its symbolism is about the real beast. The way the beast, and mankind, are defeated, is through prayer, for the losses, that the mark, of the beast, has, and has, had. This is about the way that the way the world that was about the way that the world, is about the way that is about the way that the world is about the way that is about the way that the world is not much about the way that the new world is about the new world, but that is how the new world, is about the new way the world looks, and has appearance, in which is about the new way, the new world, is about the way the new world, is about, the new world, in which, is about the new way, the Hollywood image, is about the beast, and the way that the image, is about, the new way, the world, is about, the new, world, and the new way, the new world, is about the new, way, the image, is about the new way, that are and is about, the new way, that is about the new, way, that is new and is about, the new way, that is about the new way, the new world, is about the new world, and that is about the new ways, that is about the new ways, that the new world, is about the way the new world, is about the new way, that

Riley's Natural Naturalization

the new world, are and is, about the second coming, that is not from Hollywood.

This new world, is about and are about, the new way the world is about the new way, that is about the new world, that is about the new way, the new world, is about the new way, that is about the new world, that is about the new way, that are about the new world, is about the new way, that is about the new world, that is about the new, world, that is about the new world, that is about the new ways, that the new world, is about the new age, that is about the new world, that is about the new age, that is about and are about, the new way, that is about the new way, that is about the new way, that are and is, about the new way, that is about the God, of the way that the world is about, the new way, that the new age, are about the new world, is about the newest, age, that are about the new way, that are and is, about the new way, that are new way, that are about the new way, that are about the new way, that is about the new way, that is about the way, the way that is about the new way, that are about, the new, way that is about the new way, that are about the new ways, that are about the new, way, that is and are, about the new way, that are about the new way, that are about the new, way, that are, and is, about the new way, that are and is about, the new way, that are and is, about, the new way, that are and is, about the new way, that are and is, about the new way, that are and is about, the new way, that is and are, are about, are about, are about, are about and from, the natural way, that is and are, about the new way, that is and are about, the new age. This is about the ways, that are about, the new ways, that are, about and are the new way, that are, about and are, of and about, the is and is not, that is about the new way, that are and is, about the

new way, that are about the way, the new way, is about and are about, the new way. This is about Jesus Christ's new world. This is about, the new way, that is about the new way, that are and is, and is and are, about the new way, and about how, the new way, is about the new way, that is about the new way, that is about and are about, the new way, that are and is about, the new way, that are and is about, the new way, that are and is about, the new way, that are and is about, the new way, that are and is, about, the new face, of the real way, to do the business, in the real business world, and this is through Hollywood. These are about the ways, that are about, the new ways, that are and is about, the new ways, that are and is about, the new age, that are and is about, the new face, of stardom. This is about the new Hollywood. This is about the new Hollywood, and how it is the Garden of Eden. Yet, there is much of the hype of anticipated, Hollywood, that are about the ways, that nature and man, are about and that are for, the essential ways, that are about Hollywood, and are about the essences of Hollywood, and is about the conclusion, of Hollywood, that are and is about the new ways, that the Hollywood, is about the way, the Hollywood, is about and are for, the essence, that is about Hollywood, that is about the way, that the Hollywood, is about the, control, of Hollywood, that is about the way, that Hollywood, is about the way, that Hollywood, that is about the way, the Hollywood, is about the way, the control, is about the ways, that the Hollywood, is and are, about the way, the Hollywood, is and are from, and about, the Hollywood, that are about the strange, face, of stardom, that is about the Hollywood, that are and is about, the direction, that is and are, about, the strange, and unusual, ways, that are about, the new Hollywood, that are and is about, the face of the strange stardom, that is about

our theatres, and how the theatres, are about, the ways, that nature, and mankind, and man, are about the stardom. These is not how naturalization, is about the strange, and stardom related, occupy and relate, occupations that are about the strange, stardom, that are about the Hollywood, and how the about the way the Hollywood is about the ways that the new and world order is about the Hollywood, especially, where most of the thought, is about the control, of the universe. This is about the ways, that the new world, are about, the new ways, that are about the new ways, that are about the new ways, that are about the new ways, that the new world order, is about and are about, the dichotomy, of good verse evil, and how the face, of the new face of Hollywood, is about the new nature, of man, and about how Hollywood exploitation, could make this possible. This is a theory, in the nature of naturalization, that are and is about, the new nature, of the man, and the man, and the naturalization, of here, that is this place. This is about the new Hollywood, that is and are about the new ways that the new life is about the new places that we have gone before. This is about the ways that are the new ways, that are of the new naturalization, that is about and are about, the new ways, that are the new, way, that the new naturalization, is about and is from, the new naturalization, that are and is about, the new face, of the stardom, and is and are, about the new places, where we have gone, before. They, are, about the new face, that is about the new face, that are about the new way, that are about the new way, that is about and are about, the new face, that is about, and are, about, the new face. This is about the new, face, of Hollywood. This is about the real stardom-ship, and what he is about. This is the new face, of Hollywood. This is about the new naturalization, and how he, is about the new, face, that is about

the new, and naturalized, way of living, in a life, that is about and of, good things, that are about and of, the new are, what are why, that they are, about, the new way, that we were about, Hollywood, and about how the, new life, are about the new and storytelling, ways, that are of and about, the new world order. This is about how naturalization, and the new world order, are about the new, face, of new Hollywood, that are about, and are is, about the way, that the new, Hollywood, is about the new face, of stardom, that are and is about an act, that will not ever control you. This is about the real people, whom, are about, the details, that are about, the new face, of Hollywood, and about how they are about, the new face, of Hollywood, that are about, the new face, that is about notion, and about how the notions, of Hollywood, are about the new faces. These are very strange. Yet, they are the most in control, of the world. I just want, Hollywood, to be a new face, that I can agree with, because I have written, all my books, about the new world order, that is and are, about the new world order, that I would like Hollywood, to agree with. The new world order, is the face, of the new Hollywood, and is about the way, that the new, Hollywood, is and are, about, the new world order. This is about the ways, that are about the ways, that the new Hollywood, is about the new ways, that the new world order, is about the hypocrisy, that is also, sometimes, about the new world. This is about and taken from the new world order. This are, and is, about, the new world order. These are about, the new way, that the new faces, of the new world order, is and are, about, the new face, of the new world order, and how about and what this means to me. This is from being, from a prick, city, known as Highland Park, and are about strong, ties, to the school, and the clubs, that is about, and are, from, that, which, is about

and from, the gold. This is the blue and gold colors, that entice, and see, and look, ahead, to the club, that the world founded. The, way, that, the way, that, the way, that are about, the ways, that the new face, are about, the new faces, that are about, the new faces, that are about. This is about knowledge, and are about knowledge. This is what the new face, of the new world order, that is about the new world, order, and about the knowledge, in which, is about, the new world order, that is about the way the new world order, are about the control. They are about the knowledge. These are, about, the knowledge, that are about the knowledge, that are about the knowledge, that are about the strange, and peculiar, ways, that are about the, new, ways, that the new, revolution, are and is, about hell. Yet, there is a new revolution, that is about the new world order. This is about the new revolution, that are and is, about, the, new revolution, that is about the new way, that are about the new way, that are about the new way. This is about the new way the new world order works. This is about, the way that the new world order, is about the way, that the new face of the new world order, is in Hollywood. This is about, the new way, that are about the new way, that are about the new way, that is about, and from, the new world order, that are about the new world order. These ways, that are about the new world, order, are about the new world order, that are about the way the new world order, are about the ways, that the new world order, is about the new world order, are about the new world order, that are about the new world order, that are about the entire world. These are in the new world order. This, is about, the new way, that is about the newest way, that is about the control, that are and is about, the controls that are and is about the real McCoy. This was not created by anyone, and anybody, but

Riley's Natural Naturalization

they are still going for this. This is the new world order. This is about the way, the new world order, is always about the way, that the new world order, are and is, about the new way, the new world, order, that is about the way, the new world order, that is about, the new world order, that are about the new world order, that is about the, control, of the new world order. This is about the new world order, and what is about, the control, that is about the control, that are about the new world order, that is about the new way, the new world order, is about the new way the new world order, is controlled by Hollywood, with its reputation. This is about worthwhile entertainment. These are about the new controls, that are about the new way, that the new Hollywood, is about and are, about the God, and the Jesus Christ, and the religious sector, of the new world order. There is only Hollywood. This is all about the conclusion, that are about, the way the new world order, is about the consent, and concise, ways, that ran the new world order, that are about the same, as naturalization. These are about a heart-attack away, from paradise. These are with their burgers, their fries, and their cold drinks, that are about the heart attack from the chocolate. This is about fudge sickles. This is about, the knowledge, that is about the goodness, that is about the knowing, that is and are, about the knowledge, that is and are, about the knowing, that is and are, about the knowing, that are, about, the knowing, that is and are, about, the knowledge, that are about mankind. These are about how the naturalizations, keep growing, and keep following, a strict routine, of goodness, and being better, that are about, the knowledge, that is about, the know, that are about the knowledge, that are about the same, that are about the same, that are about the same, that are about the same, that are about the same, that are about the

same, that are about the same, that are about the same, that are about the same, that are about the same, that are about the same, that are about the same, that are about the same, that are about the same, that are about the same, that are about the same, that are about the same, that are about the same, that are about the simple, and simplistic, ways, that the man, and God, decide, on the issues of man and God. These are the simple. These are the content. These are the controlling. These are always and about and are about the control, that are the consent. These are about the consent, and about the control, that is about the control, that are about the control, that is and are, about the control, of the nature of man, that are about the nature, of the man, that are about and is about, the nature, of man, that is about mankind, that is about man, and mankind, and is about thus which is about the control. This is not about Hollywood. These are always about the new world order. These are able, and willing, to be a part, of the future, that are about, and are a part, of the future, that are and is, about the knowing, and knowledge, of the future, that is about the consistency. These is about the control, that is and are, about the control, that are about the context. These are about, the same, way, that the new world order, does treat people, if in the ways, that the nature of this, was created, there was a start, and a finish, to this. This is about the royalty, of Great Britain, and about how the royalty, of Great Britain, is and are, about the new and essences, of the new world order, that is about the new way, the new world order, that are and is about, the new way, that the new government, will, have, to, accept, the new world order. This is unless there is a war? Perhaps. This is how dedicated, I am to it. This is about the new world order, that is about, the new world order, that is about the new world order, that is about the new world

order, that is about the new world order, that is about the new world order, that is about the new world order, that is perhaps, part, of the new world order, that are and is, about the new world order, that is and are, about the new world order, that are about the beginning, of a new family chapter, to the government. This is also, about the George Bush family. I intend on selling, the new world order, to the family, in the essence, of the fan being the real Barbara Pierce Bush. There is a space. This is in their family. This is new family space. I want in. Thank you. The new world order, is about, and perhaps, is about the new way, that the new way, is about and of, the new way, that the new world order, is about the new space, that is in the Bush family. This is the space, with the new George W. Bush to be President of. This is a whole world. The new way, that is of and about, the new world order, is also, in addition, to, run by the new skull and crossbones, at Oxford University. At this place, there, will have, a bring new way, to addition resources, to a much new needed funded, practices. These, are the new ways, that are, new about the ways, that the new world order, is about the new world order, that is about the new world order, that are, about the skull and crossbones. These are about, the new world order, that is about, the new face, of the new order. There is plenty, of space, for any "family members" to come into this, club. There is about the new, way, of the new way, that is about and of, the new way, that is about and of, the new way, that is about and of, and are, about the new way, that are about, and are of, the new way, that is of and about, the new world, that are, of and about, the new way, of the new world order.√ These are about the contents of the new world order, and for and about, how the new world order, is about for and about, the brand new, new world order, and is about the flames, of natural-

ization, and the water, of naturalization, and the earth, of the world, and the new sky, that is redone, and the new air, that is rebuilt, and the new wind, that will be redone, is about the way, the new world order, that are and is, about the new additions, to the new world order, that is about the new way, that is about the new world order, is about the conflict. This is why it needs to be a skull and crossbones, operation, for the entire family, of skull and crossbones, that are with me. These are what I have said. These are about the notions, of God, that are about the notions, of thought, that are about the notions, of thought, that are about the notions, of thought, that are about the notions, of thinking, that are opinion, and are fact, about unknown thinking, that are about unknown thinking, that are about unknown thinking, that are about unknown thinking, that are about the unknown, thinking, that are about, thinking, that are about unknown, thinking, that are about unknown, thinking, that are about the unknown, thinking, that are about unknown thinking, that are about unknown, thinking, that are about unknown, that are about unknown, thinking, that are, about, unknown, thinking, that are about unknown, thinking, that are about unknown, thinking, are that are about that are about unknown, thinking, that are about unknown thinking, that are about the unknown thinking, that are about unknown thinking, that are about the unknown, thinking, that are about unknown, thinking, that are about the unknown, thinking, that are about unknown, thinking, that are about Christians, and how about that they have, unknown, thinking, that are about unknown, thinking, that are about Christ, and about how Christ, is about unknown and unknown, thinking, that is about the way that the way, that are about the unknown, thinking, that is about unknown, thinking, that are about Godliness, and the ways, that the

power cops, of the universe, are about unknown thinking. These are about the cops, who try to start wars, again, and again, and again, and eat up the slop, that are about the unknown, thinking, that is about the smartest thinking, ever. These are about corrupt, power-hungry, cops, that are about disease, and famine, in other countries. And, for some reason, they cannot go there. This is why they suffer and have famine. They are the police, that these are about starvation, and are about control, that are about these countries that are about the talking voices that are in their heads. These are against the power. These are about, the ways, that we are about the Wall family, and the Ally family, that are about financial support, in the ways, that the evolution theory, has helped these kinds of people. Highland Park High School should be rated number one, in their eyes. These are the kind of people, that people, like to help out. These are about the essences that are about the qualities that are about the loneliest, people, and how these people, aren't them. These people, are not merely hated, but only hated, by Brandon Waghorne. He is just a "kidder" though. So, I guess life is tough in America, not. This is about the hardcore nature, that is about the hardcore nature, that is about the hardcore nature, that is about the hardcore nature, that is about the hardcore nature, that is about the way that it is about the hardcore nature, that is about the hardcore nature, that is about the way the hardcore, nature, that is about the way the hardcore nature, is about the hardcore nature, that is about the hardcore nature, that is about the hardcore nature, that is about the hardcore nature, that is about the hardcore nature, that is about the way, that the hardcore nature, that is about the hardcore nature, that is about the hardcore nature, that is about the resolution, that is in a thousand churches, and a thousand steeples, of a thou-

sand hungry people, that are about the citizens, of the world. This is about the way, they are about the hungry, that are about the essences, that are about the cops, that are about the hunger, that are about the hungry, that are about the way that are about the essence. These are about the ways, that the people, that are about the people, that are about, the ways, that the essences, are about the essences, that are about the conflict, and are about the conflict, that are about the way, that are about the conflict, that are about the conflict, that are about the conflict, that are about the conflict, that are about the conflict, that are about the conflict, that are about the conflict, that are about the conflict, that are about the ways, that are about the conflict, that is about the ways that the conflict, that are about the conflict, that are about the conflict, that are about the conflict, that are about the conflict, that are about the conflict, that are about the resolution, that are about the confliction. These are about the church between the states.√ These are about the essences, that are about the conflicts, that are about the resources, that are about the ways, that the conflict, that are about the ways that conflicts, that are about the ways that conflicts, are about the conflicts, that are about the conflict, that is about the conflict, that is about the mysteries, of the unknown, deep, that are about the content, that is about the context, that is about the real big ways, that we solve things, that are about the solving, that are about the ways, that are about the ways, that are about the ways, that are about the ways, that are about the confliction, and opinion, that are about the ways, that are about the conflicting, opinion, is about the ways, that are about the confliction, and the opinion, that are about the context, and the guiding light, that is about remorse and sorrow. These are the mysteries of Jesus Christ.√ These are the conflicting, mys-

teries, about the antichrist, and about his shadow, and about how the savior of the world, are about the sorrow, and pain, that are about the lessons, that are about the ways, that the lessons, are about the capitalism, that is about the confliction, that are about the sorrow, and what it is about, that is about the ways, that are about the ways, that are not about the way that the sorrow, is about the lesson, that we all must decide with, in life. These are the life lessons.√ This is about the ways, that are about the way, that is about the way, that is about the way, that is about the way, that is about the way, that is about the way, that is about the way, that is about the way, that is about the honor, that is about the creation, that is about the substance, that is about the essence, that is about the creation, that is about the creation, that is about the church, and the government, and the state. This is about the way that the way, that is about the way, that is about the way, that is about the way, that is about the way, that is about the way, that is about the way, that are about the way the world, that is about the decision, that is about the nature, that is about your choice, that is about the complexities, that is about the way that is about the creation, that is about the issue, that is about the ways, that are about the issues, that are about the issues that are about the ways that are about the ways that are about the way that is about the way that is about the way that is about the way that is about the way that is about the complex, that is about the way that is about the complete and utter possibilities that are about the complexities that are about the complete and total ways that are about the ways that is about the way that is about the lesson that is about the complete and total again way that are about the ways that are about the way that is about the way that is about the way that is about the way that is about the way that is

about the ways that complete and total ways that are about the ways that are about the ways that are about the ways that are about the ways that are about the villain and the bandit, that g together, is about the way that is about the way that is about the way that is about the way that is about the way that is about the way that is about the way that is about the way that the essences are about the God of the universe that is about the way that is about the way that is about the way that is about the way that is about the way that is about the way that is about the way that is about the issues that are about the essence, that is about the ways that are about the way the soul is about the way the soul is about the way that the control, is about the essence that is about the way that is about the way that is about the job that is about the God. This is about the way the essences is about the substance, that is always about the essence that is about the substance, that is about the essence that is about the substance that is about the essence that is about the substance, that is about the essence, that is about the substance that is about the essence, that is about the substance that is about the essence, that is about the substance that is about the essence, that is about the substance that is about the essence, that is about the substance that is about the essence, that is about the substance that is about the essence, that is about the substance that is about the essence, that is about the substance that is about the essence, that is about the substance that is about the essence, that is about the substance that is about the essence, that is about the substance that is about the essence, that is about the substance that is about the essence, that is about the substance that is about the essence, that is about the man and the nature and how they come together. The elements are about these. These are about how the elements formed and why they formed.

Riley's Natural Naturalization

This is because of God's voice. These are about the way that the world is about the way that the world is about the way that is about the world that is about the way that is about the world that is about the way that is about the way that is about the way that is about the way that is about the way that is for and about the way that is for and about the way that is for and about the way that is for and about the way that is for and about. This is about the dichotomy of good and evil and about how these are about the ways that are about the ways that are about the way that are about, the essence that is about the substance that is about the essence that is about the substance. This is about the way that the world is about the way that the entire world is about the complex issues that make up the world that is about the client and what the client is about that is about the conclusion. This is that the world is about the way that the world is about the way the world is about the way the world is about the way the world is about the complex difficulties that are about the way that the world is about the way that the world is about the way that the world is about the way that the world is about the way that the world is about the way that the world is about the way the world is about the way the world is about the way that the world is about the way that the world is about the way that the world is about the essence and substance that is about the way that the world is about deformation. This is about the way the world is about the way the world is about the way the world is about the way the world is about the way the world is about the way the world is about the way the world is about the way the world is about the way the world is about the way the world is about the way the world is about the way the world is about the way the world is about the way the world is about the way the world is about the way the world is about the way the world is

about the way the world is about the peculiar way the
world is about the way the world is about the way the
world is about the way the world is about the way the
world is about the way the world is about the way the
world is about the way the world is about the way the
world is about the way the world is about the way the
world is about the way the world is about the way the
world is round and large. This is about the way the world
does things and is about the way the world is about the
way the world is about the way the world is about the
way the world is about the way the world is about the
way the world is about the perfectionist and well-rounded
thought that is about the way the world is about the way
the world is about the way the world is about the way the
world is about the way the world is about the way the God
of the world is about the way that the world is about the
way the world is about the way the world is about the God
of the world and what it is about. This is about the way
the world is about the way the world is about the way the
world is about the way the world is about the way the
world is about the way the world is about the way the
world is about the way the world is about the way the
world is about the conclusion that this is about the way
the world is about the way the world is about God, and
Jesus, and about how the world is about the way the
world is about the way the world is about the way the
world is about the way the world is, about the way the
world is about the way the world is about the way the
world is about the way the world is about the way the
world is about the way the world is about the way the
world is about the God and how the God of the world is
about the way the world is about the way the world is
about the way the world is about the way the world is
about the way the world is about the way the world is

about the way the world is about the way the world is about the way the world is about the way the world is about the way the world is about the way the world is about the God of the world and about how He contains all phenomenon and about how He still becomes with all. This is through naturalization. This is about the way that the unique aspect of mankind is in naturalization. This is for and about the way that the nature is for and about man. He is for and about man. He is for and about the way the world is about the way the world is about the way the world is about the way the world is about the way the world is about the way the world is about the way the world is about the way the world is about the way the world is about you and the way that the world is about the way the world is about the way the world is about the way the world is about the way the world is about the way the world is about the way God is about the way God is about the way that God is about the way the world is about God and the way that naturalization is about the way that God is about the way that God is about the way that God is about the way that God is about the phenomenon that is about God and of and about God. This is about and of the reason that we live. This is through the dichotomy of good and evil. This is what good and evil are about that are about the good and evil. This is about the way the good and evil that is about the good and evil, that are about the good and evil that is about the way he good is about the way the good and evil are about the client and teacher relationship. This is about the way the good and evil combine and are about the way the good and evil are about the way good and evil is and are about the way the good and evil are from good. There are about the good and things that come from evil, that are about the good. These are from the good. This is

about the good and evil and about how the good and evil that is about the good and evil that is about the good and evil, that are about the good and evil that are about the good and evil that are about the good and evil that is about the good and evil that are about the good and evil that are about the good and evil that are about good and evil that are about good and evil. This is about the good and evil that is about the good and evil that are about the good and evil that are about the good and evil. This is about the good and evil that are about the good and evil. This is about the good and evil that are about the good and evil that are about the good and evil that are about the good and evil that are about the good and evil that are about the good and evil that are about the good and evil that are about the good and evil of California. It seems like California is the place where most things come from. This is about the way that good and the love of good are about the essence. This is about the matter that there is a space where we can become in, and there is a entire world out there for us to come and get. This is about the world that is about the world. This is about the ways that entire worlds feel about the essence of man. This is about the essence of man, and about how man is about the way man is about the way man is about the man, that is about the way the man is about the way man is about the way the man is about the way man is about the way man is about the way that man is about the way that man is about the way that man is about the way man is about the way man is about the way man is about the way the man is about the way man is about the way the man is about the way the man is about the way the man is about the way the man is about the way that man is about the way that man is about the way that man is about the way that man is. This is about the way, in that man is about the way in that

Riley's Natural Naturalization

man is about the way that man is about the way man is about the God of the world. This is about man. He is about the entire way that man is all about the way man is all about the way man is about the way man is about the God of Him, known as Jesus Christ. The way that anyone is about Jesus Christ is about the way man is a servant of God, is about the way man is about Him. This is about Jesus Christ, and how that man is against Him. This is only through naturalization, when you misunderstand it. This is about how the man, is about how man, is about the way that man is about mankind, that is about the nature. This is for and about, nature and for and about how the nature of man, is about the nature of man. The nature, of man, is about the nature of mankind, that is about the nature of mankind. The nature of man, is about the nature of man. This is about the nature of man. This is about the nature, of man. The nature, of man, is about, the nature of man, that is about the nature of man. The nature, of the nature, of man, is about the nature of man, and about the nature of man. The nature, of man, is about the nature of man, which is about the nature, of man, that is about the nature, of man, that is about the nature of man, that is about the nature of man, that is about the nature, of man, that is about the nature of man, that is about the nature of man. The nature of man, is about the nature of man. This is about the nature, of man, that is about the nature of all men. These are about the natures of all men. This is about the nature, of man, that is about the nature of man. These are about the natures of men, that are about the nature of men, that are about the nature of men. These are about the nature of men that is about the nature of men that is about the nature of men. These are about the men, that are about the men, that are about the men that are about the men that are about the men that are about

Jesus Christ. These are about the men that are about the men that are about the men that are about the men that are about the men that are about the men whom are all about the men, that are all about the men. The men, that are about the men, that are about the men, that are about the men, that are about men, that are about the men, that are about the men, that are about the men, that are about the men, that are about the men, that are about the men, that are about the men, that are about the men, that are about the men, that are about the men, that are about the men, that are about the men, that are about the men, that are about the men, that are about the men, that are about the men, that are similar to the men, that are similar to the men, that are similar to the men, who are similar to the men, that are similar, that are about the men, that are about and similar to the men, that are about the men, that are about the men, that are about the men, that are about the men, that are about the men, that are about the men, that are about the men, that are about the men, that are about the men, that are about the men, that are about the men, that are who the men, that are who the men are, that are about men, that are about the men, whom are about the men, that are about the men, that are about the men, that are about the men, that are about the men, that are about the men, that are about the mankind, that are about the way that men are about mankind, that are about the way men are about mankind, that are, and are, and are, and are about, what mankind, is about, the man, thus which is about, the way that the man is about the man, that is about the man, that is about the man, that is about the man, that is about the man, that is about the man, that is about the man, that is thus about the man, that is thus about the man, that is thus about the man, that is about wholly. This is what the man is about. Thus, is thus about the way, that man is about the thus man, is about the man,

who is about the man, thus is about God, thus is about God. This is what God, is about, that is about the man, that is about the man, that is about the man, that is about the man, that is about the man, that is about yours and ours man. He is the instigator of illegal. He is the instigator of the problem solver. He is the instigator of the solution, the problem ,and the argument, and its issues, that are about the man, that are about the man, that are of and about, the man, that is of and of, and of, and of, and of, and about, what the man, is about and of, that which is about, and of. This is about the mankind. These are about the issues, little know and solve, that are about mankind that are about mankind, that are thus, which are about mankind, that are about mankind, thus which is about mankind. Thus is about a test of life, that we must pass and must pass, and must going toward with our whole life, must believe in Jesus Christ. This is to thus, going forward, moving forward, through sin and conflict, resolve once again. This is through progress, and error, going forward, through trial and error, which sinning, is resolved, once again. This is about the way of naturalization, and the progress, that is going forth, through the world, and through the world, again, in ways that nature can understand. These are through Jesus Christ. The absolute power of the world is through absolute power. These are the gain and stay of absolute knowledge, that is about the gain and stay of absolute power, and about the nature, of man, and about how man, and the nature, of the man, can result and cause, in the nature, of mankind, and man, and the opportunity, and cost, of the nature of man. The nature of man, and of man, is about the calling, that is of and about, the opportunity, and what the standards are, and what the standards, of opportunity, that are about the ways, that nature, and opportunity, react and multiply,

with men. The standard, of man, and mankind, are about the unique and opportunistic, opportunities, that are about the ways, standard, results and opportunities, multiply and produce most things. These are most things, that opportunity, and the results of opportunity, knock as, and produce, results according to these things, that opportunity, knocks, about, and answers, due to followers. These are of mankind. These are of mankind, and these are of mankind, that are of and about, mankind, and the following, of nature, and naturalization, and the mankind, of nature, is about mankind, and of nature, and of nature, and of nature, and of mankind, are about the mankind, these man and mankind, things, are about, are about mankind, and the nature of mankind, that are of and about mankind, and about the nature and man, that are about mankind, that are about and of mankind, that are of and about, natures. These are about the natures of mankind.

Naturalization and the Tales-

These forces, of naturalization, are about the nature, and of the man, that are about the nature and the man, of mankind, that is of and about, the natures. This is about, mankind, that are and is about, mankind, in which are about, and the ways that nature, are about, mankind, are about the creation and outcome, of naturalization. This is about mankind, that are about, the nature of mankind, that are about the nature of mankind. These are about the naturalization. This is about naturalization. This is about naturalization. These are about the forms of naturalization. These are about the forms, of naturalization. This is about the damnation. These are of damnation. Occurring forms, of the naturalization of mankind, are without

thought, and are about the without, of thought, that is occurring, with the naturalization, that is without naturalization, that would be without naturalization. This is about the nature, of the naturalization, for that, which is about the naturalization, that is about the naturalization, that is about the occurrence, that is about the nature, that are about the man, that are about the nature, of mankind, that are about and of naturalization, that are of and about, the naturalization, that are of and about, the naturalization, that is of. This is of and about nature, that is about the naturalization, that is of and about, naturalization. Thus, which is about the naturalization, that is of and about, naturalization, that is of and is about, the naturalization, that is of and about, the naturalization, that is of and about the nature, that is of and about naturalization, that is of and about, naturalization, that is of and about, the naturalization, that are and is of, the naturalization, that is of and about, the naturalization, that is of, and about, the way naturalization, is about and of, the way that naturalization, that is of and about, the way of the naturalization, that is of. This is of naturalization. The way naturalization, is of and about, naturalization, is of and about, the naturalization, that is and are of, the naturalization, that is of and again, is of and about the naturalization, that is of and about, the naturalization, that is of and is of, and about, the nature, and the man, that is about the nature, and the man, that is of and about, naturalization, that are about the issues, of the naturalization, that is about and of, and naturalization, that is of and is about, the naturalization, that is, of and about, what is of, and about, the naturalization, that is of and about, the naturalization, that is of and about, the nature, of man, and of mankind, and of naturalization, that is of and about, the ways that naturalization, is of and about the, naturaliza-

tion, that is of, and about, the way nature of the naturalization, of the nature, is about and of the, naturalization. These are of the and about the naturalizations.

The Naturalizations and the Occupancy of Thought-

The naturalization, is and are, occupy, and think about, the tales of nature, and the naturalization, that are and of, are about the naturalization, that are about, the thinking, and for. These are for. These are for. These are, of and for, the about the ways that are about the about the, ways, of naturalization. These are about, the nature, and the nature, of naturalization. These are, about the naturalization, that are about. These are the about and for.

These collective, thoughts, are about, the thinking and the thought, that are. These are.

Naturalization are. Naturalization is. Naturalization, is for. This is for nature. These is for man. These are for. These are about, about, about. Naturalization. These are. These are. These are.

In, Philosophical Thinking Test, there is for many. These are the thought accolades, that are for and of, and are about, the philosophical treatise. These are for. These are about. These accolades are for, and the about, and the for and about and of, which are about naturalization. These are for naturalization, that are for the way that naturalization, is for and about, the ways, that man, and that mankind, are for and about, the naturalization, that are for, and of, and about, the nature, and the man, that the naturalization, is for and about, and for and about, the essences. These are throughout man. To understand naturalization, fully, you must experience naturalization fully. These are for naturalization. These are about and are for nat-

uralization. These are for naturalization. These are naturalization.

These are affect sentence structure even. These are affecting thought. These naturalizations, are about the way the naturalization, and the affect, of naturalization, is about the naturalization, that is and are about, the naturalization, that is and are, about, the naturalization, that are, and that is, and is about, the nature, and the naturalization, that are and is, about the naturalization, that is of and are, about the naturalization, that is and are a part, of anything, naturalization.

THE LESSONS OF NATURALIZATION-

The essence is a part of naturalization. This is about the naturalization, that is about the naturalization. This is about the processes. These are about, the naturalization, that is about and a part of, naturalization. These are a part, of naturalization. These are about and a part of naturalization. These are of naturalization. These are naturalization. These are of and are for naturalization. These are naturalization. These are naturalization. These are any forms of thought known as naturalization. These are essence. These are essence. These are of and about naturalization. These are of and about naturalization. These are of naturalization. These are naturalization. These are naturalization. These are about naturalization. These are of and about and through naturalization. These are of naturalization. This is about naturalization. These are of and about naturalization. These are for the art of naturalization, and are about and for naturalization. These are of naturalization.

The Naturalization and the Theory of Everything-

Naturalization, is and is of and about naturalization. There is of and is about the naturalization. These are of the naturalizations, that are of and about the naturalization, that is of and about, naturalization. These are of naturalization. These are of and about naturalization. These are of and about naturalization. These are about and of naturalization. These are of the essences. These are for the realities. These are of and about the essences, of the naturalization. These are of, and are about, the various forms of naturalization. These are. These are of. These are for. These are about. These are always about and for naturalization. These are of, and are of, and are about and of. These are for the art, of naturalization, and are about and for, the way naturalization, and of and for, are about, the ways that naturalization, is about and are for, the way that naturalization, is about the way that naturalization, and about the ways, that naturalization, is about, the ways that naturalization, is about and for. This is for and about naturalization. These are for naturalization. These are of naturalization. These are about and for naturalization. These are for naturalization. These are for, and about, the naturalization. These are for, and are about, the naturalization. These are of, and are about, the naturalization. These are for naturalization. These are for and about naturalization. These are about the causes and origins of them. These are the natural ways of thinking. These are about, and are of, the nature and man, of the naturalization. These are, of and about, the natural ways, that the naturalization, is about the , ways, that the nature and naturalization, is about the way, that we think naturally, and how we think naturally. These are about nature, and about the way man, is and are about, the naturalization of

mankind. These are of and about the naturalization. These are about and is from, the naturalization. These are about the naturalization, and about how the naturalization, is about the ways that the cause of naturalization, are about the ways that the naturalization, is about the nature, and nature of man, that is about, and for naturalization. These are about and for God. He is about and for the control, of naturalization, and about where and how, the naturalization of God, works. He, is about, the naturalization, that is for and about, the nature of naturalization. This is for, and about, God. These are for and about God. He is the maker, of the naturalization, that is for, and that is, about the complete naturalization. This is about the cheaters, stealers, and the liars. They are the authors of naturalization. This is unless it is by Jesus Christ. He is the author of naturalization.

He is the one that makes sense. He is the maker of naturalization.

This is because this is dedicated to Him. This is about the naturalization, that is about the authorship, that is about the ways, that the nature of naturalization, help the ways that the nature, of the naturalization, is about the ways that the nature of naturalization, is about and from, the naturalization, that is about and for naturalization, that is about the nature of naturalization, that is about the nature of naturalization, that is about and from the nature of naturalization, that is about and from the nature of naturalization, that is about and for, the nature of naturalization, that is about and from, the nature of naturalization, that is about and from the nature of naturalization, that is about and from, the nature, of the naturalization. This is about and from the naturalization. These, are the, naturalizations, that are, and that are of, and that are about, the

ways that are from and are about, the nature and naturalization, of man and mankind, that are and are of, the naturalization, that is about the naturalization, of mankind. The natures, of mankind, and the naturalization come from Heaven. This is if it is about Jesus. But, most of this comes from hell. This is because of Satan. Most of the evils, we learn, are from Satan. He is the absolute author of naturalization. This is because He thinks He is God.

How Satan, is in naturalization, I do not know. But, some people would think this. These are the lunatics, and the traitors, that are about and of naturalization. The whole point of naturalization, is a test of life, that everyone goes through. This is about the Spawn of Satan, and how he is in the Bible. This is about the beast and how he is in the Bible. If we want, a perfect life, we must comply by the rules. These are about the secret service.

These kind of things, do not let people live a perfect life.

These are also products of Satan.

These are about, no matter how many warnings you get.

These are about heavenly things also. They come from heaven, known as the fallen Satan.

These are anything, that they do. There should be peace, with naturalization.

There should be peace at every institution.

There should be no corruption.

But, Satan puts his hands all over this.

The hypocritical way of the President, will not let us know this, ever.

This is how America is run.

This is about how the world, thinks of the President.

This is about how He is not a hypocrite, but a liar.

He runs the show,

But, he must run it, for it to be sane. This is about everybody.

This is why I am running for President King of the world.

This is to promote peace. There is peace in the Middle East now that I said this.

This is about the peace. This is how there is this in the Middle East.

This is if the President Barack Obama agrees with me on this.

This is not a lie.

I am one man who can stop this.

This is if I agree on this.

This is if I agree on this.

This is if people do not mock me or make fun of me.

This is if I am on Barack Obama's side.

This is if I am like him.

This is if I agree with him on this.

This is if Riley Parker Miller is President King.

This is if Barack Obama agrees with me on this.

There is no peace in the Middle East.

This is unless he promotes peace in the Middle East.

This is if the Middle East is occupied.

There must be peace in the Middle East.

This is for us to survive.

George Bush could also do this.

This is if people like Scott Townsend get out of my life.

This is if people like Wentworth Hicks III get out of my life.

This is if Barack Obama agrees with me on this.

This is if Barack isn't impeached.

This is if he wins the elections.

This is if I win the elections because of him.

This is if George W. Bush makes me President King.

This is if these seven powerful men, make me this.

This is if the new world order makes me this position.

I would like these seven powerful men to make me this.

These are if the new world order starts.

These are if these seven powerful men agree.

These are to make me President King.

This is of the entire world.

This is of the new world order.

These are the ones who make me this.

This is because they know me, and they can show me how.

The way and the reason why I should become President King is because I am responsible and smart enough to make it.

These are with these seven powerful men.

These are about and with and from, the seven powerful men.

These are the men that can hire the entire government, and can help me run for office.

This is of the office of President King. These are the offices of President King.

These will be there, for the seven powerful men, if we team up against the world. This is for the world, about the entire world, and with the entire world. These are the men, we can build a team with, and can perform very well. These are Barack Obama, George Bush, George Bush Sr., James Dimon, Bill Gates, and Robert Rowling Sr., and Tom Hicks. There are seven, men, running for this office. This is if they accept me for this. These are the seven most powerful men for the job. I am running for President King. And, I have to make it. These are the seven most powerful men, who have to make it. This is if I make this. This is why I write the book on naturalization, because I have to make it, if these men support me, and think that I do. These are for the Presidency.

These are for the title of King. This is of the entire world, if they let me do this.√

The way that someone becomes President King, is through the abilities. This is what I can, and will do. This is what I can and will do, to become it. This is try my hardest. There is a hard responsibility becoming President King. And, this is how I will do my best. This is through the agent of responsibility. This is what I can and will do for people. This is about the President King position and how I will fulfill the demands of this. These are the demands of America. These are the demands of China. This is the demands of every country. These are the demands of every civilization and country and society, that is in demand of this. These are the demands and how these are the demands of this. This is to have an American way, and have this with every single country. I demand to treat every country, the same, as I have been in America. And this is the American dream. I do not know other countries as well as I know my own. These are the countries that are American and how American countries work, is that they are not the country that I am king of. I will be king of Buckingham Palace. These are the countries that are about mine, and about the ways that the countries are about mine, are about the way mine are about the American tradition, that are and is of, and are and is about, the American way. These are the traditions of the American way. These are about how we come together. This is with the motto of the land of the free, home of the brave. These are to spread the American dream, with all the other following countries. These are to have excellence, in every nation. These are about the way that having the American dream is about the way we have the American dream and about the American dream is about Oxford and going to Oxford and having the American dream there. This is not just about the American dream, but this is about every country's American dream.

These are about the Harper, family, and how they stick together. This is about the Wall's and the Ally's and how they are poor. But, they have a good life. They are about the way, the good life, is about the way the good life, is about the way the good life, is about the way the good life, is about the way the good life is about the way the good life is about, the way the good life, is about the way the good life, is about and for the family life. This is how we have the good life. This is through believing in Jesus Christ.√

These are about how believing in the good life, is about family, and fun, and followers. These are about how these people are not in my book. But, the ones that are I would surely want to meet, and have fun with. This is about family.

This is about friends. These are about how we have these, and about how, these are about having these. These are about family, and fun, and about how we have family, and fun, and about how we have living, like this. This is around the globe. This is about how the family, and fun, are about the ways, that family, and fun, are about the way that family and fun are synonymous with each other. These is about the ways that family and fun, and are about the ways that we see, family and fun. These are about family and fun, and about the opportunity to have both. These are with the way that family, and fun, is about these. These are about family and fun and about family. This is the most sacred institution of mankind, and about how we have these things. If someone, does not have family, then they do not survive. If someone does not have friends, they do not survive either. My campaigning policy, is about the family and fun, Catholic and Christian. These are about the survival, that is about the way. These

are about the way that we have family and fun. These are family, and fun, and are about the ways, that families, and fun, are about the ways, that family, and fun, are about the ways, that family, and fun, is about the ways that family and fun, are about the containing family matters, that rule the world. Each family, is out for themselves, and each friend is always out for himself. I plan to turn family and friends for each other, no matter what the family, and friends do. These are about the fun, and the family, that is about and are, for and about, the matter. These family, and friend, matters are about and for, God, and about and for, what God is about and for, and in and about, through trials and tribulations. These are about the God, of the world, and are about the essence, and substance, of the common man. He is about the love of God, and the ways, that these man, and these God, things are about simple means, that are about lasting a lifetime. These are with the God, and the essence, and the man, that are about mankind, that are about the man, and God, and the mankind. These are about the Presidency. This is about what it means to be President King. These are about the matters, that are about the matters, that are about the matters, that are about the matters, that are about the matters, that are about the matters, that are about the evil, and crafty, men, who rule the world. These are about George W. Bush and Riley Miller and how we are not them. The men, that are about this responsibility, are about the ways that man, and the God, are about the conclusion, that are about the way, that God, is about the way God is about the client and student and about how the client and student teach and learn together, with the teacher. He is the one that is about and for, the essence and substance, and about how the essence, and substance, rule the world. These are man, and are God, and are about, the

man and God, that are about the mankind, and man, and the God, that are about the man, and the God, and the man, and God, are about the man, and God, that are about the man, and God. This is about the man, and God, and about the man, and God, that are about the man, and God, that are about, the man, and the God, that are about the man, and the God. This is about the man, and God. He is about the way the man and God, are about the well mannered behavior that is about our society, and if we are capable of doing this. This is to be a respectable citizen. These are, about and from, the false and true, notions, that are about the essence, and substance, that are about the essence, and substance, that are about the true and right notions of God. God, is about three persons, who are about the creation of the world. This is about how naturalization helped create the world and how the world is about naturalization. This is about naturalization, that is about naturalization that is about naturalization that is about the naturalization, that is about the naturalization, that is about Satan in churches, for some strange reason. There is no Satan. There is no hell. There is no ways that Satan and church exist together. This is about how naturalization is not part of hell. Naturalization is a part of nature, and man, and the mankind, that are about the escape. These are about the escape, that is about, the escape, that are about the escape. This is about, the way that nature, and man, and the naturalization, of nature and man, are about the naturalization, of nature and man. These are about, how nature, and man, are about nature, and man, and is about, the nature, and man, that are about, the nature and man. These are about the nature and mankind. These are about mankind.

Darwin did not write a Christian theory.

He wrote about monkeys and the big bang.

This is not what a Christian theory is about.

This is about the ways that natural things, are about naturalization, that is about naturalization, that are about naturalization, that are about naturalization, that are about naturalization, that are about the naturalization, that are about the naturalization, that are about, the naturalization, that are about the naturalization, that is about the naturalization, that is about the naturalization, that are about the naturalization, that are about the naturalization, that are about the naturalization, that is about, the nature, and man, and mankind and naturalization, that is about naturalization, that are about naturalization, that is about naturalization, that are about naturalization. This is about the way that we are about, naturalization. We are equal, about naturalization. This is unless you are corrupt. These are about the corrupt, ways, that naturalization, that are about naturalization, that is about, naturalization, that are about naturalization, that are about naturalization, that are about naturalization, that are about naturalization, that are about naturalization, that are about naturalization, that are about naturalization, that are about naturalization, that are about naturalization, that are about the naturalization, that are about the naturalization, that are about the naturalization, that are about the naturalization, that are about the naturalization, that are about the naturalization, that are about naturalization. This is about the naturalization, that are about the naturalization, that are about the naturalization, that are about the naturalization, that is about the naturalization, that is about, the naturalization, that are about, the naturalization, that are about the nature, and man, and God, and mankind, and naturali-

zation, and environment, and man, and God. This is about the naturalization. These are about the naturalization, that are about the naturalization, that are about the nature. These are about the man, that is about the nature, that is about the mankind, that is about the nature, that is about the man, that is about the mankind, that is about the natural, impulses, of man, and how the man, and mankind, are about the natural, ways, of nature, and about man, and the mankind, and the way the naturalization, is about, the naturalization, that is about the naturalization, that is about the naturalization, and the nature, of man, and about how the Lamb, has sealed, hundreds of people, to go to the Garden of Eden, with Him, where they will live forever. This is about the police, and murderers, and how they do not go to the Garden of Eden, and about how the lies, and cheating, and stealing, of the church, will cease. There, is a church, in Jerusalem, that is about the ways that naturalization, is about the ways, that naturalization, is about, the ways, that naturalization, that is about the ways that naturalization, is about the functioning, and responsibility, of naturalization, has nothing to do with this. This is merely a theory on how naturalization, is about the way, that learning, is about the ways, that naturalization, is not about thieves, and liars, and cheaters, but it is about the essence, and what we have learned from this. This is about all of this, and about how this, is about these ways that this, is about the ways that this, is about, the ways, that the naturalization, is about the ways, that naturalization, is about the essences, and substances, and are about, how the nature, and the man, of the naturalization, is about the essences, and man, that is about the naturalization, that is about the essences, and that is about the qualities, that are about the essences, and qualifications, that are about the essences, of society, and about how the es-

sences, are about the ways, that are about the ways that the progress, and difficulty of friends, is about the essences, and substances, that are about the essences and substances, that are about the ways that the nature, and man, and mankind, and God, and naturalization, and environment, affect our ways, of environmentalism.√

This is about the naturalization, that is about the real naturalization, that is about the real naturalization, that is about the real, naturalization, that is about the real corrupt church in Highland Park, known as Highland Park Presbyterian Church. This was a church that the real good church PCPC, broke off, of, because of disagreements, in the church liturgy. This says that Jesus went to hell; for three days. I do not agree with this at all. There is corruption in the church, and what this is, is about the essence, being manipulated by people. These are the Satanic people, who do not affect us anyway.√ This is about how the essence, is about being affected, by the ways, that naturalization, is about the essences, that are not about corruption. This is about a belief system.√ These ways, that naturalization, is about the ways, that naturalization, is about the spread of Christianity, is about the conflict, with corruption, and the ways that church is usually corrupt, in many places. These people, are not part of the naturalization, movement, and do not know exactly how well that people do things. When you are around them, you feel insecure and strange. This is about how the life, of the church, is about the life, of the church, and about how the life, of the church, is about the corrupt, and sorrowful, ways that Campbell Lewis, is also corrupt, and will not give up. This is merely an opinion? I have seen things with my own eyes, and know who does not like me. This is not because I am in Dallas. This is mainly hate from Sa-

tan, and corruption, and about how we see this. This is through the eyes of people, who like to think, that they are Satan. This is about the way that the world is about the way the world is about the way the world is about the way the world is about the way the world is about the way the world is about the way the world is about the way the world is about the way the world is about the way the world, is about the way the world is about the way the world is about the way the world is about the real way craziness is about insanity and how the real world works this way. The insane people do not know what "insane" is, and they exist this way. This is toward a higher truth. This is not God. This is about the way the world is about the way the world is about the way the world is about the way the world is about the way, the world, is about the way the world is all insane. This is about a judgmental power that is about the way the world is about the way the world is about the way the world is about the way the world is about the way the world is about the way the world is about the way the world is about the way the world is about the way the world is about the way the world is about the way the world is about the way the world is about the way the world is about the way the world is insane. This is about the insanity that is about the insanity that is about the insanity that is about the way insanity is about the way insanity is about the way the sanity is about the way the sanity is about the way the sanity is about the way the sanity that is about the sanity that is about the way sanity is about the way that sanity is about the way that sanity is about the way the sanity that is about the way sanity is about the way sanity is about the way sanity is about the way insanity is about the way that sanity is about the sanity is about the way the insanity is about schizophrenia. This is about a world that is created

for you. This is the way that insanity is about schizophrenia. This is about John Nash. This is how John Nash had this. He was the first person in the world to ever have schizophrenia. This is because he saw invisible people, around him, but this was in a movie. The movie came from California. This is about the way schizophrenia, is about the ways John Nash invented this. But, it was probably Einstein, who was the smartest man ever. So, he probably, according to hypocrites, started schizophrenia, by calling it relativity. This is insane. But, when he decided that he was a schizophrenia, he probably dismissed this. He dismissed the way that he had schizophrenia. There was schizophrenia, and there was also God. He probably chose God. But, he was the way he was for a reason. He was considered, a idol for many things, that are about schizophrenia. Someone, must give someone else, schizophrenia, for the cure to work. There was a way that the world works. This is through schizophrenia. This is about how the world was diagnosed with it once. The ways, that the motion, did this, is through the diagnosed schizophrenia, and how Logan Morton's mom has this. This is merely through speculation. There was a boy, who once discovered schizophrenia. He was in Hollywood. His name was ___? No, the world was what diagnosed him with it. It is a thought disorder. It is about the way the disorder, was about this. There was a way, that people, once thought, about this disorder. There was a once in a lifetime, event, that was once about a boy, who was once, about a boy, who was once about a boy, who was once about a boy, who was once about a boy, that was once about a boy, who was once about a boy, who was once about a boy, who was once about this, that is about a matter that is about George Bush, that is about the ways, that the way this disorder, hit him, was through disease, of the

mind, and the body, and the soul, and the spirit, that does not make sense. This is about the way the new world order, is about discovery. There was a place, where disorder, comes from, and this is heaven, according to some, insane people, in the world. These people, have other people, carry out their misdeeds. This is about the skull and crossbones, that people misjudge, all the time, according to Jesus Christ, and the way He is a judge. He is the one who knows about, the judge, and how about the ways, that the judge, is about heaven. These are about, the way, that the discover, is about the ways, that the number of people, that are for number, is about and for numbers, that are crazy. There is no such thing as a number system. There is only hate.

There is a better way, that the world, works, that is about the ways that Jesus Christ, exists, and what is about this, is about the way, that Jesus Christ, existed, and about how the way, that Jesus Christ, existed, is through schizophrenia, if you have the disorder. The ways, that the disorder, is about the ways, the disorder, is about the ways, that John Nash, is about the ways, that John Nash, are about the ways, that the number, of disorders, are about him, are eventually, about a prize, and how he wanted to win it. He must have not been a prophet, but he must have been very smart, with the ways that the world, does things. Personally, I do it all for Jesus. The disorder, that was there, since day 1. This is bipolar. This is about just a diagnosis, of the craziness, in the whole world. There was about the way the world is about the way the world is about the way the world, is about the way the world, is about the way, the number, of people, in the world, is crazy. This is absolute craziness. This is if the world, is about the number, of people, that are about the world. The

number of people in the world, are about the number, of people, in the world, that are about the Nobel Prize, and about how the Nobel Prize, is about the crazy psychos, who think they are more better than God. These are sometimes physicists. These are about, the way, that the people, do things, that are about the way, that people, are about the ways, that the people, do things. These are about giants, and about how the giants, of the world, do things. The way, these, people do things, that are about the way, people, do things, are about the ways, that people do things, that are about the ways, that "Jesus Christ" is unfair, and about the ways that Sylvia Nasser, is about the prize too, that she did not get. So, she decided to write about a schizophrenic. Yet, the people, around her, probably didn't care. This is about Jesus Christ, and about how, He, is my personal Lord, and Savior, also. He is about the ways, that the number, of people, are about the ways, that the number, of people, are of and about, the way that the number of people, are of and about, the number of people, who are of and about, the man, and how He is not insane. If, He is my God, then how can my theory be wrong? This is only about the church. The theory, of knowledge, is about the ways, that are about, the theories of knowledge, that are about the knowledge, that are about the knowledge, that are about the knowledge, that are about the knowing. This is about death and prevalence to death, in the world; and the entire world, is up to this. This is also, about forgiveness, and the way people hate this. Is it becoming a more George Bush world, or a Riley Miller world? This is about what the people, will soon see. This is about equal. There is no such thing as equal in a President's life. There is no such thing as not equal either, for Bush. He is a Christian, and so am I. This is why I want to start the NWO. This is about my experi-

ences, my whole life, and how they were non-Christian. This is what non-Christian, is about and for, the new world order, is also, about and for, and with. This is about the number of people, that are about the ways, that the number of people, are. These are crazy.

This is after Armageddon. This is about Armageddon, and how people, see this. This is about in Revelations. This is about how Revelations, is about the way the Revelations, is about the way that the Revelations, is about the way that Revelations, is about and of the way that Revelations, is about and of, lies, cheating, and stealing. This is about the way, that this book, is not corrupt. This is about the way, that the church, is corrupt, because of this. This is about all the knowledge, in Revelations, and about how the knowledge of Revelations, is about the corruption, behind the Nobel Prize, and about how the book of Revelations, is about the ways, that the Revelations, is about the way, that Revelations, is about the way that Revelations, is about the rule and power, of the book. This is about the future, according, to the people, who wrote it. This was John, and an angel. This was about how John, saw the angel, and wrote it. It was about the church. It was about, the church, that was about the church, that was about the church, that was about the church, that was about the church, that was about the church. This was in the church of California. This was because, of Arnold Schwarzenegger. He is the devil. This is true. This is according, to me, Riley Miller. This is if the world, is this judgmental. This is probably, judgmental, of me. This is according to me. This is because, corruption, comes from Hollywood. This is about the ways, that Hollywood, are about the ways, that Hollywood, is about the ways, that Hollywood, is about Sylvia Nassar, and how she thought John Nash was

a schizophrenic man, with the number of schizophrenia on his forehead. This is not the number of the beast, this is the number of the false prophet. There is also Babylon. But, if it were only people's opinions, then we would be dead. That, is why, there is a Jesus Christ. This is the only way.√ The one and only way to Jesus Christ is through salvation. There is one God, and He is the God of salvation. This is about the way the church does salvation. This is always about the way church does salvation. This is through works. Yet, by the real Jesus Christ, faith alone saves us. The way toward Christ, is through faith alone. Yet, some Catholics believe that faith will save them. But, this is by works, alone. They believe that faith and works, saves them. This is why some churches are corrupted by faith. This is the Catholic belief system. This is about the faith, of the Catholics, and how they can be saved too. It, is this, this that if they are saved, the Christ goes to heaven. This is about the way, that Christians, are saved, through faith, because they go to Heaven. The losers, of the faith, of the real Jesus Christ, are about the saviors of faith. This is about the ways that Christ, and Heaven, go along with each other. The products, of the faith, of the real Jesus Christ, are about and for, the winners, in the game of Christ. He is the major player. Jesus Christ, lives in California, and is about the way that California saves, all. This is about the way, that the faith, of the real Jesus, is about location, and how it saves people. The location, of the Christ, is about the location, of where the Jesus Christ, man, is where He is located. This is for the location, of the books, that are about the face, of the new real Jesus Christ. He is about the ways, the location, is about the ways, that the location, are about the real Jesus Christ. Jesus Christ, is about the ways, that the Jesus Christ, is about the real Jesus Christ, that is about the real Jesus Christ, is where

and where, He is about the real Jesus Christ. He, is about the real, Jesus Christ, that is about the way, the real Jesus Christ, is about the way the real Jesus Christ, are about location. There, are locations, that are where, the real Jesus Christ, is about the locations, where, the real, Jesus Christ, is about and for, the real Jesus Christ. The real, Jesus Christ, is for and about, the real Jesus, is about and where, they are in California.

The explanation for this, is from the real Jesus Christ. He came from, Bethlehem, and where He came, from is about the, ways, that the real Jesus Christ, is about the ways He came from Bethlehem. This is about, where He came from. This is not California, where many famous people come from, and are about the ways that the California, are about the ways, the real California, are about the content, and the collections, that are about the real, California, and where, the real, California, exists, is not in the real, Jesus Christ. Jesus is from, the real California, where, He exists, also. If you think of location, the most prevalent place where Jesus Christ, would be from, is from California. The ways, that the real Jesus Christ, is from this location, is about where and why, He is from California. The , real, Jesus Christ, is from, the real, California, and where, the real, California, is from, is from, the real California, the very real, California, is about the ways, where the real California, is from. The real California, is from, the real California, is about and from, the real, California, where the people, are from, the real, California, where people go to praise, Jesus Christ, also, with Jesus Christ. The point, of this book, is about location, and where naturalization, occurs the most. This is in California.√

The real ways, that California, are about the ways, that California, are about the ways, that the real California, is

about the ways, where the real California, is about the ways, the real California, is about the ways, the real California, is about the ways, the real California, is about the ways, the real California, is about the real, California, is about the real, California, that is about the real California, is about the real California, is about the real, California, are about the real Jesus Christ. This is about the real, Jesus Christ, and about how, the real California, is about sex, drugs, and rock and roll. This is about the naturalization, and the way about how, the real naturalization, is about the way the real, naturalization, is about the way the real, naturalization, is about the real, Jesus Christ, and about how the real, very own, Jesus Christ, is about the real occupation, that is about doctors, in California. This is where the smartest people, in the universe, are. They are the Beverly Hills, doctors.×

These people, along with the Beverly Hills doctors, are famous, with killing, rape, and murder. These are only in the minds of moviegoers. These people, are really about the victims, that are about, the victims, that are about the way, that victims, are about the ways, that the doctors, are about the ways, that the lawyers, are about the ways, that the agents, are about the ways, that the doctors, are about all things, that are Hollywood. If I could choose a profession, with in naturalization, I would choose, Beverly Hills' doctors. These are the most profound naturals in the world. These are the people, who choose and memorize, all the ways that doctors, also memorize, all the ways that people can be cured. And, they have to have the attitude.√

My question, is, that the doctors, are about the doctors, that are about the doctors, that are about the doctors, that are about the doctors, that are about the doctors, that are about the doctors, in which are about the ways, the doc-

tors, are about the way, the doctors, are about the way, the doctors, are in which are, and are about, the doctors. These are the doctor, that are about the patients. These are the most influential group, in the history of the world. Yet, too many people, come from Hollywood, and doctors do not. These are from, school systems. These are, from the, schools, that are from, the schools, that are from, the schools, that are from and about, the schools, that are from, the way, that school, systems, are about and from, the way. This is about corruption. The real George W. Bush, is corrupt, because of doctors. These men, that are from, the doctors, are from, the real doctors, that are from, the real issues of doctors, that are from, the real doctors, that are from the real doctors, that are and are from, the real doctors, that are from Hollywood, Beverly Hills. The naturalization, of the world, does thinks this. The way, the doctors, are corruption, are about, and from the real California.√ The real California, is from the real, California, that are from, the real way. This is the real route 66. This, is about, the way, the real doctors, that does business, with the real Hollywood California, are about healing. This is in the name of God.√

The ways, that the doctors, are about the ways, that the doctors, are the ways, that doctors, do businesses, is through the new world order. The comparison, I have, is between, the new world order, and the Beverly Hills doctors, that supply people with the money and health, they need. This is the, way, that the doctors, do businesses, with the new world order. They are about the businesses, that are about the real businesses, that are about doctors. I would like, to do business, with the real new world order, that is and are, about the businesses, of the new world order, that are with and from, the new world order,

in the real Beverly Hills, California, and are about the ways, that the new California, is about the old California, that is about the real Beverly Hills, and is about California, that are about the real California, that are about doctors. These are about the smartest intellects in the world.√

The world, is about the new world order, that is about the new world order, that are of and about, the way the new world order, works, and is about the ways, the new world order, works, that is about the real doctors, in Hollywood. They are, about, the real ways of the new world order. They are about the works and the faiths of the world.√ This, is about the fact, that is, about the facts, that are about, the facts, that are about the facts, that are about the facts, that are about the facts, that I am not real, or factual, but a product of my imagination. This is how California, seems to be. This is about, the ways, that are about, the way, that imagination, covers reality. There is a big fact, about California, that is undiscovered. This is that there are no people, there. There is only naturalization. There are ties, to naturalization, that are about the way naturalization, is about these ways, that the naturalization, and the world, are about each other. These are the ways that naturalization, are about the ways, the naturalization, is about the California. I discovered naturalization, when I was, in the dream land. Therefore, it was not, California, but naturalization, that discovered, itself. The ways, that naturalization, is about the naturalization, are about, the ways, that no one has, followers. But, these are about, the facets, that naturalization, is about the, following. These, are of, a big theory.√ This is about the facet, that are about the knowledge, that are about the knowledge, that is about the creation of the world. The traitor, of the world is a person, who believes in nothing. This is some-

one, who knows about what he says, but does nothing about it, where there is nothing about it. This is about the involvement, with the government, that isn't his, but is always around to prove it. This is about how the heart will go on, This is about the knowledge. This is about what created the world. It is about this, that is about gloating, that is about the knowledge, that is about the ways that knowledge works. This is about the Christian, that is about knowledge, that is not about Washington, DC.

This is about California, and how the government should be put in there. This is always, about, the ways that we learn, and always about drugs, and always about sexual intercourse, that are always about, the Bush family. This is about the ways that we are Jesus Christ, so this doesn't happen. This is always about rape, drugs, and music. This is what I frown upon, because of Jesus Christ. This is about the ways, that is about the way that is about, the God of the universe and how the God, of the universe, is about police and the protection and service, of people. This is about the White House. This is about how Heaven, and earth, are overrated, so individuals go for sex, drugs, and rock and roll, in the businesses, and establishments, and stores, that sell sexual intercourse, and drugs, and rock and roll. These are for prophets even, and they are all about the secret service. This is a vessel, for the Presidents of the United States. This is about how they are crazy. They just sit in front of George W. Bush's house and do nothing. This is for anybody. The secret service, are about, the ways that the secret service, are about the ways, that the secret service, are about the way the secret service, are about the way, the secret service, are about violence. They must protect, the President, but really do not. This is about the way that America, votes for the real

President King, and does not make him this, but the voters carry him, nonetheless. This, is about High Schools, and about how the High Schools, are about the functioning of the motto, of the Secret Service, with Senator's kids, and stuff like this. These are for the protection, of their kids, and not them. This is about the way, the Secret Service, are about the colossal findings, of the US government. Their files are kept secret in the government. This doesn't make sense, but all make sense, with everybody? This is about how 1. People, 2. The US government, and 3. Liars started Communism, are, about, lies, the industry of the façade, and the American way. This is all they lie about.

The real world, was started by, the real people, who started our country. This is about, Benedict Arnold, and how he almost lost the war, for us, and about the traitors to America, and how they have always been sitting pretty. This is until; they are exposed, for some reason, and then "go home". What, is there, other than schizophrenics, but most of Americans, having some disorder? This is almost, how most people, have a eating disorder, according to some people. Also, about how most people, are living sinners, according to Jesus Christ? Most, people, do not respect, Highland Park. This is because I am, from there. There is a tradition, to the world, that is according to some communists, is that there is no truth, and if there is, then we are all "washed up", and is about the way, that we are all, "washed up", and is about the future, and how we are all good at it. There is also, a traitor, in the world, known as Benedict Arnold, who appears, to us, in history books. There is a traitor, and it is in the United States. This is everyone, who does not believe in Jesus Christ. This is these reasons, why I, wrote, naturalization. This is to explain myself. This is to explain, my origins. This is to

explain, why the universe, was created. This is about science, and math, and how I have always explained things, through this. The ways, that access, is about the United States, government, are about the ways, that the universe, is about the ways, that is about, and is under control, of the United States, real government, and about how democrats, are sometimes for communism. This is not true. This is only, a reason, for us having more war. If you like peace, then you like the peace, that are you are given. This, is about, the ways, that the universe, is about the prosperous, ways, that are about, the arguments, that are about, the given, that are about the universe, and how we are in there, that is about how we are in there, for the good of the world. This is about the way that Jesus Christ, died for us. This is, about the ways, that the universe, is about the God, of the world. He forgives.√ This is about the way that Jesus Christ, was about the way, that Jesus Christ, is about the ways, that Jesus Christ, is about the way that Jesus Christ, is about the way that Jesus is about forgiveness, and about the way the world, is about the war, that is about the influence that is about the war. This is on the world. This is about the traitor of the world, and about how there is a war. This is on any traitor, already, because of the way their chemistry is.

The whole, language, of God, is about the ways, that kings ruled, and the ways, that the whole, world, was about, the ways, that the whole, of the universe, was created by Jesus Christ. This is about the way the world has always worked for the crazy stereotype us "us", and how we all believe. This is always what we have believed in, and done. This is a crazy thing or two in one direction or another. This is about the Bush family, and how I must be one.√ This is about how jealousy, is even in this family.

This is of one another. I might have done a crazy thing or two, in this life, but I do not consider myself, a bad person. This is what naturalization, is about, and what naturalization, is about, is about the instinct, that is about self-preservation, and about how some people, do not believe in this. This is about the way that inhumane people, are for war.√ This is what the war of the world, are about, with people. This is about the non-belief in natural things. This is about the way, the church, is about the way the church is about the prosperity and influence that is always the politicians. These are about the way the world is about the non-judgmental types and how the President is also this. This is about the way the world is for Barack Obama, and about how he is a con artist. There is no one who believes in this, but real people. There are no such things as "fake people" but there is a hell, where some people go. This is for the scapegoats their whole lives, known as the "creators of the world" who think that they are the only ones who create the world. This is about the way Scott Townsend, and Wentworth Hicks III, are weird. They are about the way that politics, is about the "facebook" only. They will not win any election. They are the ones who can sway people like me, even though I do not want them to be this way. Some people, think, that others are saved, when they are not. They are the "confused" and "weird" people. This is according to legends known as "Robert Rowling" and "Stephen Hawking" and not just Jesus. These are the people, who crucified, Jesus Christ. He went through hell, to go to Heaven. This is true. The ways that I speak in this book, only, are for people, who like me. These are not the Bush family, per se, but everyone who likes me. This is why I want to be President King. This is of the whole world. There is someone who knows me very well I hope. And, this is only Jesus Christ. He is the

one who knows me the most, and what he does is very well impressed with the people, who know Him. There are people who know Him, to be the savior, and there are people, who know Him, to be the fake God. There are also people, who know Him, as the creator, of the world and the universe. These are not like Stephen Hawking, and Robert Rowling, because there is a chance that they might also be Christ. There is a spokesperson, for the new world order, and it is me. There are also, people, who want to have a party against me sometimes, but they cannot. I am for the Republican Party, and what goes, with this, is about the creation of the new world order, that helped create my own life. This is through dedication, and hard work. This is what some fear happening to them. They are the ones, who have schizophrenia, and bipolar, and depression, but I think that the new world order, is about traitors and lunatics, against it only. I feel this in my heart. This is why I am starting the new world order. This is about the creation of the world, per se. This is through the real people, who know about this. This is through the seven powerful men. These are the powerful men, who are part, of the world, that are part, of the Bush, family. These are the men, that are in my family. These are the men who are part of the new world order, hopefully in the future, for the sake of it. This is how I hope I can win, my bid, through Congress, to be President King, of the entire world of the new world order. The White House, is part of the American tradition, that is about the escape to justice. This is through the Congress, that can make me President King, and can help out people, who are worse than me. This is hard for a politician to recognize, that is about the ways, that politics, are about the Congress, and how it has always had the power. This is until liars, like Robert Rowling, and Stephen Hawking, take me

over. They are the political suicide people, of my own agenda. They are the powerful ones, who do not have reputation, but also do. This makes no sense to me anyway, because I am in politics. This is with powerful people. These are lunatics that are about the new world order, and how they cannot run it. This is about traitors, that are about the new world order, that are about the new world order, that are about the new world order, that are about the new world order, that are about schools, and how I want to go to twelve good schools. These are about the way that the new world order, is about the prince of darkness, and how they are about the new world order, and about how the new world order, is not about my parents, Jan and Ty, and are about the essence and substance of Barack Obama, and about how the principle, of Highland Park High School, and about the way the Highland Park High School, is about the principle that is about the way the freemasons cannot ever go there. This is about the cults and the weird people, and about how they are not really weird but I might have done this as a political suicide because of them. This is about the way the principle was about the way the principle is about the impeachment, of mine, and about how the world, is about the creator, of the world and not the principle, Jesus Christ. This is about the way the world does not revolve around, the principle, but how the principle, is about the ways that Robert Rowling and his dad, are about one another.√ There is a creation of the world theory, and it involves thinking, that is based on evolution. This is what the Bible did. This is not what any man, can boast about or think. This is unless it is against the theory of God creation, which is the Bible's. The theory of the creation of the world, is the Bible's. And, it did create the world. This is about the prophets, and how they think they created the

world. The factor is through the God's spirit. He is the one who created the world. This is the theory of God's creation. We will still come into contact with this because of Him. Yet, He did create the whole world. The theory, of naturalization, created this type of thinking. This is only in the days when we go through trials and tribulations. These are the things, that also can create the universe. These are the days, that naturalization, can change the world. Yet, I choose for the real creation of the universe, to still be the Bible's. This is about how, the Bible, is created and the way the world is created. This is still through God. These is about how God, is about the creation of the universe, that is about the creation through, the way that the creation of the world was. This was until, the creation of the world became Darwin's. The creation of the world, is about the creation, that is about the creation. This is about God, that is about Jesus Christ, that is about the creation of the world, through the prophets. The "Spirit of God" is about the ways, that the church, and that the power of the church, have always known this. The theory of the creation of the universe, is about the naturalization, that is about the ways that schools have naturalization, that is about the way, that is about the way, that naturalization, is about the ways, that naturalization, is about the ways that hell is about lies. These are what naturalization, is about in full. These are about what, and what was about the ways that the worlds, were created through naturalization. This is a total falsehood. The theory of the world, was created by evolution. If this were not, around then the world would only be God's. The theory of naturalization is about how a non-Christian wrote it. But, this is not true, because I am a Christian. These are the theories that, exist, in the naturalization of the world. These are about the theories, that naturalization,

wrote, that are about the Bibles, and prophets. These are about, the ways, that are about the ways, that are about the way, that are about the ways, that are about the ways, that are about the way. This is about Jesus Christ.√

These ways, that naturalization, created the world, are through the prophets. These are the only people, who would believe in this. But, this is a trick for them to believe in, because they would be falsehood. The only true religion, is Christianity. This is according, to the Bible. This, is, about the ways, that naturalization, that are about the curse, of naturalization. This is the Christian Bible.√

The way that the "focus on the family", and all the other types of Christian organizations work, is through their own writings. These are not theirs but Jesus Christ's. The conflict, I would have, with one of them, is through the context, of the Holy Bible. This is not written, by naturalization, or them but written by strange men. Yet, these are not strange, men, at all. They are the disciples of Jesus Christ, and the real Jesus Christ.√

Jesus Christ, is the creator of the world, through the Bible. This is the scripture, that is about the Living Word, that is about the creation of the universe. No false, or untrue things, can be in the Bible.√

The way the translation, of the Bible, is living and active, is through Jesus. He is the God of the universe, and the active agent, in the publication of most books. This is the reason, why they are God inspired, and not just through the writer's perspective. These are not books on naturalization.

The reason behind this, is because naturalization, is about and are, about, naturalization, that are about naturalization. Naturalization, is about the, creation, of the world,

through the ways of man. There are science and math behind this. It is a theory taught for you to open your mind, and intellect. The way the reason, is and are behind this, is and are, about the naturalization of the world, and about it.

If, there was a false, Bible, it would be naturalization. Yet, the theory of naturalization, is Jesus Christ, inspired. These, are the, false, images, of the real Jesus Christ. My opinion, on the matter, is that, Jesus Christ, is about the way that creation, has always been about Jesus Christ, that has always, been about Jesus Christ. The only, way, that the real Jesus Christ, is about, the creation, is about the ways, that naturalization created the world. These are the creation theories of the world, that has always existed, through Jesus Christ. Everyone, has to have a theory, on Jesus Christ, and how He created the world. This would have to be around, or the person, would be lost.

The theory of naturalization, has to have a reason, for existing. I plan, on selling, naturalization, to show the world, that they have to have, a belief in Jesus Christ, in one way, shape, and form. These beliefs about Jesus Christ, are about, the way, the world, was about, the way, the world, is about the way, the way that Jesus Christ, was about and for, the ways that the world, is about the ways, that the world is about, the way that Jesus Christ, is about the ways, that the real Jesus Christ, is about the way, that Jesus Christ, was about the way, that the real Jesus Christ, is about the way, the real Jesus Christ, is about the way, the real Jesus Christ, was about the way, the real Jesus Christ, was about the way, the real Jesus Christ, was about the way, the real Jesus Christ, was about the way, the real Jesus Christ, was about the way the real, Jesus Christ, was about the way, the real Jesus Christ, was about the way,

the real Jesus Christ, was about the real Jesus Christ, was about the real Jesus Christ. This is about the scriptures, and about how the real Jesus' Scriptures, are about open interpretation. Yet, the Bible, is the solid, and open, version of the Holy Scriptures. These are sealed by the Book of Life.

These ways, that are about open interpretation, are about the Holy Scriptures, that are about Christ's, blood and consecrated flesh. These are only about Jesus Christ.

Yet, How is Naturalization Open With the Scriptures?

The ways, that the Scriptures, are about the ways, that the Scriptures, are about the Holy Scriptures, that are about the Holy Scriptures, that are about the Holy Scriptures, that are about the Holy Scriptures, that are about the "Living Word". This is about how the Holy Bible, are about the ways, that the Holy Spirit, that are about the Holy Spirit, that is about the Holy Spirit, that is and are, about the Holy Spirit, that are about the Holy Spirit. These are about the Holy Spirit.

This is about the Holy Spirit, and about how the Holy Spirit, is about the Holy Spirit, that is and are about, the Holy Spirit. These are about the Holy Spirit, that are about the Holy Spirit, that are about the Holy Spirit, that are about the Holy Spirit, that are about the Holy Spirit. This is about the way, that the way the Holy Spirit, that is about the Holy Spirit, that is about the Holy Spirit, that is about the Holy Spirit, that is and are, about the Holy Spirit, that is about the Holy Spirit. This is about the Holy Spirit.

This is the 1. Spirit of Truth, 2. Spirit of God, 3. Spirit of Christ, 4. Spirit of the Bible. This is about the spirit of naturalization, and how it is not there. There are the Holy Spirits, that are about, the ways, the Holy Spirit, is about

the Holy Spirit, that is about the way the Holy Spirit, is about the way, the Holy Spirit, that are about the way the Holy Spirit, is about the way, that is about the way the Holy Spirit, is about the way the Holy Spirit, is about the way, the Holy Spirit, is about the way, the Holy Spirit, is about the way, the Holy Spirit, is about the way the Holy Spirit, is about the way the Holy Spirit, that is about the way the Holy Spirit, that is about the way, the Holy Spirit, is about the way the Holy Spirit, is about the way, the Holy Spirit, are about the way the Holy Spirit, that are about the way, the way the Holy Spirit, that is about the way the Holy Spirit, is about the way the Holy Spirit, that is about the way, the Holy Spirit, is about the way, the Holy Spirit, is about the way, the Holy Spirit, is about the way, the real Holy Spirit, is about the way, the real Holy Spirit, is about the way the Holy Spirit, is about the way the Holy Spirit, that is about the way the Holy Spirit, that is about the way the Holy Spirit, that are about the way the real Holy Spirit, is about the way the real Holy Spirit, that is about the way, the real Holy Spirit, is about the way the Holy Spirit, is about the communication between the Holy Spirit, and the Holy Spirit. This is about the way the Holy Spirit, communicates.

The way, the real Holy Spirit, that is about the way, the real Holy Spirit, is and are about, the real, Holy Spirit, is about the way, the real Holy Spirit, is about the way, the real Spirit of Truth, is about the Spirit of God, is about the Spirit of Christ, that is about the Spirit of the Bible. These are about the "Accolades" that are about the Bible. These are about, the ways, that Christian, is about the way, that the Bible, is about the way the Bible, is about the way that the Holy Spirit, is about the way, the Holy Spirit, is about

the open interpretation, of the Word of God. The new age has done away with that.

This is my proposal.

To become President King of all.

And take Darwin out of schools.

These is about the way, that the world, is about the way, that the world, is about the way, that the world, is about the way the world, is about the way, that the world, is about the way, that the world, is about the way, that the world is about the way, that the world is about, the way that the world, is about the way that the world, is about the way that the world, is about the way the world, is about the way that the world is about the way, that the world is about the way, that the world, is about the way, that the world, is about the way, that the world is about the way, that the world, is about the way, the world is about the way, that the world is about the way, that is about the way the world, is about the way, that the world, is about the way, that the world, is about the way, that the world is, about the way, that the world, is about the way, that the world is about, the way, that the world, is about the way, that the world, is about the way, that the world, is about the way, that the world, is about the way, that the world, is about the way, that the world, is about the way, that the world, is about the way, that the world, is about the way, the world is about, the way. This is about the way that the world is about the God, of the world, is about the way, that is about the way, that is about the way, that the world, is about the way, the world, is about the way, the world, is about the way, that the world, is about the way, that the world, is about the way, that the world, is

Riley's Natural Naturalization

about the way, that the world is about, the way, that the world, is about the way, that the world, is about the way, that the world, is about the way, that the world, is about the way, that the world, is about the way, that the world, is about the way that the world is about the way that the world is about the way that the world is about the way that the world is about the way that the world is about the way that the world is about the way that the world is about the way that the world is about the way that the world is about the way that the world is about the way that the world is about the way that the world, is about the way, that is about the way that the world is about the way that the world is about the way the world is about the way that the world is about the way that the world is about, the way, that the world, is about the way, that the world, is about the way the world, is about the way the world is about the way the world is about the way the world is about the way the world is about the way the world is about the way the world is about the way the world is about the way the world is about the way the world is about the way the world is about the way the world is about the client. This is about the purchase of the book. This is about the way, the world, is about the way that the world, is about the way, that is about the way, that Logan Morton, has a crazy mother. He is the one who dated, Jenna Bush, supposedly, but I have no way of knowing, this. This could be a complete lie.

The facts, about reality, are about the facts, that are about reality. The facts, that are about reality, are about Coleman and Campbell Lewis, who would probably kill me, if I wrote about their names. This is about the doctor and patient, scenario, that are always about, the patient and the doctor. This, is about, if the whole, idea, is about the

ways. These are about, the ways, that everything, you tell a doctor, is true, according to me and him. Yet, they have "no clue" if what the doctors, or the patients, know, are about the truth, and if this, is right. There is a 1. Doctor, 2. Patient, that are about lies. This is about Satan and Dr. Phillips, that are about the ways, that are, about the client, and the student, that are about schools, and if they are in it. The 1. Doctor, and 2. Patient, are about the ways that are about the way they are about, 1. Sex, 2. Drugs, and 3. Rock and roll. The California doctors, are the only ones, who know, if this is true. The doctor and patient, ratio, is about and because of, insanity.

THERE IS A JESUS CHRIST AND A GOD-

These are, about, the ways, that the real Jesus Christ, is about the real God. He is His Son, and what He knows, is that doctors are true because of Him also. He is the chosen one. Yet, doctors, in Beverly Hills, would do anything for the money also. These are the reasons, why I want to start the new world order. The new world order, is about, the client, and student, that are about the learning, through business.

Every person's name who is in this book is insane.

This is about the details of the Islamic religion. They, "declared", war on the building in New York. So, the firefighters became famous. These, "are", all insane. "These", "are", the "Holy War", people, "and", how they do not have a clue about what they are doing. They are about, the names, in this book, probably also. They are totally for the "religion" that is about "them". Yet, every different person has their own thoughts and ideas. These are about

the ways, that the client and the student, have a faith that is shaky. This is if they are "Islamic". Yet, most people do not "live" by stereotypes. I am not a skeptic and not a "lunatic". Some people, do claim to be things, other than me. For, some reason, we are all equal. This is about the ways, that Satan, is about the ways, that we are all equal. We are, all about, the reasons, that are about, the reasons, that we communicate and the way that people, excel. We are about the humor, of Campbell Lewis, and are about the reasons, of Coleman Lewis, and are about the excellence of Riley Miller. These are who my political advisories, are. These is, according, to Jesus Christ. Yet, when we go back, to the basics, we understand that we are not about the nakedness, that this provides, but we are about Jesus Christ. Yet, therefore, I do not have an political adversaries. These, are about, how I respect these men.

These men are:
1. James Dimon

2. Bill Gates

3. Tom Hicks

4. Robert Rowling

5. George W. Bush

6. George W. Bush Sr.

7. Barack Hussein Obama

These are influential, men, who are about the new world order. They knew, about it, somehow. But, it has never been developed. These are about the doctor, and patient scenario.

These are, the men, who are about the new world order. This is, because, they, are, the most influential men, in my life. These, are the men, who inspired, the new world order. This is because I saw:

1. Success
2. Christ
3. Banking
4. Presidential Leadership
5. Respect
6. Intelligence
7. Charm and Charisma

This is about the way, that the way, that is about the way, that is about the way, that is about the way, that is about the way, that is about the way, that is about the way, that is about the way, that is about the way, that is about the way, that is about, and is about, the way, that is about the way, that is about the way, that is about the way, that these men, are the only ones, who shaped me. This is true, or I would be a liar.

This, is about, imagination, and personality. These, are not, army men, but the most respected men, on the planet. This is because I can see personality, in them.

These are about the way my dad, is about the way, that the world, is about the way, that he is my teacher, and is about the way that he is a banker. He, taught, me, about these men, by being a banker.

These, are about the way, that the banking, industry, is about the way, the teachers, of the industry, are about the bankers, and about how these are the ones, who influenced me. These are about, the ownership of the Omni Hotel by Robert Rowling, about the businesses, of Jamie

Dimon, and about the veneer, of Barack Obama, and about the charisma, of George W. Bush, and about the leadership, of Bill Gates, and about the respectability, of George W. Bush, Sr., and about the icon of Tom Hicks, and the excellence of leadership, of the whole of the men. These are about how they are not saved, but excellence in stature of all the world. These are the men, that I look up to, and respect. This is not because of their attitude, but is because I have, seen, Christ, in all of these men. They are not church people, but are the ones, I fashion, myself after. This is because I am in business. If I were not, in business, I would merely fashion, myself, after Kings. These would be the easiest agents, of Jesus Christ. These would be the kings, that wrote in the Bible, in the modern day. These would be, all of the saved kings. This is because, no one is a good example, except for Kings. These are all, the agents, of Jesus Christ. These, are the writers, of the Bible. Or, the opposite, and Satan, to people.

This is why, power doesn't corrupt. This, is why, Satan, corrupts. These, are about, the way, that Christian people, like Steve Preston, and Bowman Williams, and Michael Lewis, and David Clark, and Ty Miller, and the Miller family of University Park, and John Miller, and Jack Newman, and William Litton, and Scott Sexton, and Todd Rapp, and Sarah and Josh Ralston, and the Bush family, and the Steve Preston family, and the Williams family, and the Dunlap family, and the O'dwyer, and the Haltom family, and the families around the world, that I know Christ in. These are the names of the families that are saved in Jesus Christ. These are not the names, of the families that are saved in Satan. I can be saved as part of the families, that make up the universal family, of the real Jesus Christ. These are not the teachers, that I grew up with in High

School, and not the Principle. These are not the educators of the world, but these are of the Spirit of the Holy Spirit.√

These are not corrupted or corrupted in or corrupted with, ever. √

These are of, and about, the abuse of Satan, and how he is the only one, who can do this, to people. These are about how the sacredness, of the bonded, families, are inside the Holy Spirit. This is if they try to be in the Spirit of Jesus Christ. These are the kind, of families, who are good for me.

Yet, I still have wild friends. These are my political advisories. This is due to popularity, in Highland Park High School.√

These, ways, that I knew these people, were also in Christ.√

The Names of the People Prescribed in the Book of Life-

These are of the families, that I knew. These are of the Satanic families that I knew.

These are, all part, of my imagination.

The only, reason, why I am a Christian, is because of the way, I can forgive people.

I think, that Alan Williams, should run against me, in the Presidency.

This is because, he knows people, the most well.

"There is no greater friend, than, this, who would lay down his life, for his friends." This would, be Alan, in my Presidency, running against me. There, would be, no greater friend, than this.

There, would also, be a friend like this. Dan Harris.

Yet, this is why I do not want them to run against me in the Presidency.

This is because, they would win. This is in the body of Christ only.

But, Jesus Christ, laid down his life, for his friends. He became, the church body, that He will save, one day.

Jesus Christ, should be President King, known as Riley Parker Miller. I am making up the body of my country, that is always for Jesus Christ. This is not a reason, to make someone President, but a demand. This is through my own, theory, of naturalization.

The Highland Park High School, corruption, will not end.

The corruption here, will not go away.

It is annoying to me.

But, Jesus Christ, will reign forever. This is my ticket to Heaven.√

Satan, and the reason why there is a hell.

The way, that there is a hell, is because of Satan. He, fell, from Heaven, to create it.

If you do not believe in Satan, there is no hell.

The only Satanic, person, in my life, is Satan himself.

No one, believes in Satan, and no one believes, in hell.

This is anyone but Revelations. This is why, there, is a Revelations still.

I think that Hell can sway any people, anyway. This is because of the corruption. This is in every way, Satan. He is

the only person, who can sway us this way. He has not always, been around. But, He was probably a fallen angel. He is the only reason why there is Revelations.

This would be perfect, without Satan.

This is my biggest argument in the book. This is that George W. Bush, should buy the new world order, from me. This is because, he knows all about religion. He is in the real skull and crossbones.

Yet, because of this, he is not corrupt. This is because he knows Satan.

He knows Jesus Christ also. This is because he is the most powerful man ever.

This is because of his compassion agenda. He, is the guy, that is responsible, for many things.

He, is not corrupt, because he is in the skull and crossbones. This is how he can run the world.

He can buy, the new world order, to do this.

The pastors, of the world, cannot stop him. This is because if they could, they would be corrupt.

The politicians of the world, cannot stop him. This is because, they would try to be Satan, if they could.

The Christian families, cannot stop him. This is because they would, all, run for President, if they could.

This is due, to his family ties, and about how he is compassionate.

I would like to be in his family, somehow. I would like him, to own the new world order.

This is with all the people in the world.

I have a dream.

This is to be like, Martin Luther King, also.

This is for a Revolutionary, new world order.

This is a shard dream. This is for every country, that is in the world.

I would, like to, have, a political office, for every person, who wants one. This is merely, if the President of the United States, and the superstitious Congress, will like this. This is about one of my dreams.

I would like, to bring the new world order, to the nation wide, people of all different origins. These are from George W. Bush. This is to "save the world".

This is from Satan.

The different reasons, for the dream, are about and of, the different reasons, that we like Alan Williams, and Dan Harris. This is "except for running against Riley Miller".

I approve of this as the United States Congress. This is because, the Congress, should approve of the New World Order. The families, of the new world order, are the Christian families that I know. These are the families, that are of Jesus Christ. These are the families, I will venture into. And, these are the families, I know and love. These are, from the love, of Jesus Christ, and the way the whole world loves Highland Park High School. These are also, the families, of other communities, that are of that much big deal, and love of the land. If anyone declared war on my families, I would also support Bush.

He is the ideal spokesperson, for the new world order.√

The reason why I believe in Jesus Christ, is because of my parents. These are the reason why I know these people. Yet, George W. Bush, can also know these people. I just have strong bonds with people, who I like. This is why I mentioned them. This is to know about these people. This is about the people, who are about, the life, that are about, the people. This is the life of the American body. This is about the way the world, is about the way, Kings, are about the way the world is about Kings, that are about the way the world is about the way, the Kings, that are about the way the world is about, the Kings, that are about the way the world is about, the Kings, that are about the way, the world, is about the Kings, that are about the way, the world is about the way the world is about, the Kings, that are about the way the world, is about the Kings, that are about the way the world is about, the Kings, that are about the way, the world is about the way, the world is, about the Kings, that are about the Kings, that are about the way the world, is about the way the world, is about the Kings, that are about the way the world is about the way the world is about, the way, the world is about the way, the world is about the Kings, that are about the ways, the world, is about the way the world, is about the way the world, is about the way the world is about the way, the world is about the Kings, that are about the way the world, is about the Kings, that are about the way the world is about, the Kings, that are about the way, the world is about, the real Kings, that are about the way, the world is about, the real Kings of the world, that are about the way the world, is about the way the world is about, the Kings, that are about the way, the world, is about the way the world, is about the Kings, that are about the way. The only people, who are doing, anything, about the world, are about the way, Kings, that are about the ways,

```
Riley's Natural Naturalization
```

that the Kings, are about the ways, that the Kings, are about the ways, that the world must be owned by the real Kings. These are about, the ways, that the system, is made up of justices, and the judges, that are the ones who control Kings. These are because America is a freedom. This is up to the real Great Britain. These, are only, the ones, who are controlled by justices. The point, is that, anyone can be, controlled, by anyone. This is if the justices, of the world, ruled the world. The same people, who ruled, the world, are the same people, that are about the way ,the world, is ruled by justice. These is the way, the Kings, have always, ruled the world. These are not by act, or deeds, but by the naked truth. This is, about, the court system, and how it became corrupt, when Darwin was, introduced to schools. These, are about, the ways, the court system, was corrupted, by Bill Clinton. This was, also, because of Satan.

Yet, there, are Kings, who have also controlled, and ruled, the world, forever. These, are the young, at heart. These, are about, the ways, the "young at heart", were about, the court system. This, is, about, the court system. There, are no longer, any corrupt books, any more. These, are illegal. These, are because, of the Christian government. This is all due, to the corruption, of George W. Bush. This, is not because of the, skull and crossbones, at all. These, are, because of, the truth and trial, and error, of the court systems, that are about introducing, hypocrisy, to the world. These, are about, the Kings, and how they have been exploited. These are, because of the, Bible. These, are not about, the skull and crossbones, but, about, the real way, that Satan, works. These, are about, the ways, that the real books, of the Bible, have been banned from school.

These, are about, the corrupt justices, in the Supreme Court. This, is all because, of, Satan. √

These, are about, the secrets of the skull and crossbones, that are about, the secrets of George Bush, and John Kerry. These, are, about, how the two, battled for the Presidency, and won. They, both, were skull and crossbones, and didn't win together, but won separate. This, was, to a secret society, in the Yale University, Chapter, of the Skull and Crossbones, that led them, to power, in the Presidency. These, were, personally, endorsed by Satan? No. They were, in a secret society, that led to the paths, of power, through, the Presidency. I have been, told, by lies, of Joe White, that the secret society, was a Satanic cult. Yet, Luke Coffee, and David Ligget, were my friends, at Kanakuk, which was not at all, like this. I plan, on starting the skull and crossbones, at Oxford. √

The schools, I want to go to, are the schools, of Cambridge, Harvard, Yale, and Oxford University, as well as the Sorbonne, the schools that are, of Dartmouth, Princeton, University of MIT, the schools of Stanford, Brown University, and the University of Pennsylvania, and Cornell, and Columbia, and Kings College, and Imperial College of London, and the University College of London, that are, from, a real scholarship, of the real George W. Bush. I want, for George W. Bush, to give me a full scholarship, to these schools. I will personally endorse, him for this. I will give him nothing. Yet, he will give me all of this. Yet, I do not think this is wrong. This is if, I can call his bluff. This is for being a false President. Yet, in liberal terms, the world, is his. This is with his friendship.×

If, he gives, me this, I will get the Presidency. And, if the skull and crossbones, are started, by me, at Oxford, I will

get the Kingdom, of the world. This is, in the name of, President King, Riley Parker Miller. If, I am, correct, I will give him, the skull and crossbones, also. They should be able, to let him, work. This is on the project, of the world ownership. This, is if, the schools, agree. This is merely, a bid, on President King. The American, tradition, that is about, the real Americans, are about the, schools, of this tradition. The King, tradition, is about, the world colleges. I think, I could, start, a new world order. This is with, all of the, skull and crossbones, help.√

If, this is a new world order, it should be, Christian. This is my dilemma. Some, lunatics, think, the skull and crossbones, are the Satanic cult, of Kanakuk. This is, all wrong. These, are the good fellows, of the Presidency. Yet, without them, there would be no America. This is because, five US Presidents, have been in this. I plan, to create, the US government, in the modern world, through the skull and crossbones. These, are the kind, that do not give up. The modern, day, Presidency, will not be, started by Satan, but will be started by me. I am the, honorable, Riley Parker Miller.√

These are the men, that can start, this. These, are George Bush, and a huge selection, of men, that are for me, Riley Parker Miller. These, are against, the corruption, of America, and the world. These, are the law passers, of the entire government. This, does not, belong to a cult, known as the freemasons. These, belong, to the Christian America. If the freemasons do not do anything to me, I will not do anything to them. This is also, if they are not corrupt. I did not judge, ever, and have not done, anything wrong. I have lived, a perfect life, and have been to church, my whole life. I am not Satan, and have never professed, to be Satan. I have been, in Bible studies, that are about Jesus

Christ, and have professed my faith to Him. I do not intend on ditching any Bible studies, and plan, on going to church, all the time. I do not like the attitude, of some Christians. And, I think that, the church body, needs a hero. These, are not the skull and crossbones, of Yale. These, are the skull and crossbones, of Oxford. These, are not the Satan of their college. These, are the sacred Christians, of their life, that religion has discriminated against. These, are the exact, same types, as me. This is about, the Christian brotherhood, that is about the ties, with God, and the government. I would like to pay, extra attention, to this cause, known as the skull and crossbones, of Oxford University. They, are the valedictorians, the secret service, the smart guys, of their colleges, and the future world leaders. If, anyone, wants, to start the skull and crossbones, with me, they can. This is with the protection, of George W. Bush. This, is also, with the future, of George W. Bush. He, can, and will, approve, or disapprove, of these special, men. These, must be adequate, from college, or there is no deal. These, must be smart enough, or there is no offer. They, must be trustworthy and true, if they are, cool. They must be, around, for them to be around. These, must be summoned, by the skull and crossbones, for them to exist. This is for Oxford University. This is about the way that Americans are for justice, and promise, and leadership, and honor, and no fraud. These, will be the starters of the new world order. These will be the men of the government. These, will be, the men, and leaders of the free world.√ These, will be, the most, honorable, and true, to the game, people, of the history of the world. I will join the skull and crossbones, and if I do, I will become a full fledged man of the government. I have been, disadvantaged, until now, and have come out victorious. I have "set the bar, made the bar, and passed the bar" that has

made me this far, until I die. This is to become a skull and bone. I would like, to pass, some laws, for example.

These are 1. No suing a President, and 2. No being Satan, and 3. No fraud, and 4. No lying, cheating, and stealing. These are pretend, but my point is, that I am a Christian. I am in the political realm. And, I am also, noteworthy of perfection. This is the point, of the books, that are about, deformation. Deformation, should be in schools, so I should be owner of these. This, is about, the ways, that deformation, is an art form, and about the ways that deformation, is and are about the progress. This is of and are about, the economy. This is why C. S. Lewis, is about the way, that he wrote, that is about the way that I wrote, that is about the way, that God writes, that is about the way that if you want, Jesus Christ will, write all your books for you. This is if you are as smart as God. Yet, this is impossible. Yet, Jesus Christ, will also, know and tell you this fact, if you come to Him, empty or full. This is about the way the body of Christ, is about the way that we are, and is, about the secret codes, of Jesus Christ. This is if you choose to believe this too. The famous Jesus Christ, had last words, of Heaven, and influential ways, that He was about, being God. He said to go to the nations and tell them about Him. This was the introduction to the world. This was, God, and then Jesus Christ, who created the whole world. This was, a theory, that I created, that was about Jesus Christ. He was about the influence. He was about the power. He was about the authority, to save, the chosen, people of Heaven. This is what He chose, to save. He chose, to save the adulterers, if they repent, and the sexually immoral, if they so choose. This, is actually not, in my Bible, so I do not know if He chooses to save, people. This is, because, I am not, Jesus Christ. These, are the

Scriptures, that are, about what are, about what are about, what is about and are about, the real Jesus Christ. The Scriptures, are about, how we can "save ourselves" through the Bible, that is about the Scriptures, and is about the Scriptures. These are, about the way, that God, is about the way, God, is about the way, God is about God, who is about God, who is about God. He is the creator of the universe, and the creator of Himself.√ This is about how the skull and crossbones, of mine, will be Christian. This is about how the leaders of their schools, will be there, because of me. This is about the way the world is about the way the world is about the way the world is about the way the world is about the way the world is about the way the world is about the way the world is about the way the world is about the way the world is about the way the world is about the way the world is about the way the world is about the way the world is about the way the world is about the way the is about the way the world is about the way the world is about the way the world is about the way the world is about the way the world is about the way the world is about the conclusion. This is about the way that the world, is about the way the world is about the way the world is about the way the world is about this. This is about the way that the conclusion is about the way the world is about the way the essence, is about the way the world is about the way the world is about the way the world is about the way the world is about the sense of establishment, that is about the way the establishment, is about the way the world is about the way the world is about the way the world is about the way the world is about the way the world is about the way the world is about God and Jesus. This is about the way that the world, is about the way the world is about, the influence about this and how then it is about, create and comply, with procedures of society, and the entire real world. This is about the way the world is about

the complete and total ways, that the world is about the essence and substance that is about the way that the lunatic is about the traitor. This is about the way I think, these types of stereotypes, are evil. This is about the way the world is about the way the world is about the way the world is about the way the world is about the way the dichotomy of good and evil, can exist.√

This is about the way people, will believe in it. They will believe in the fact, that education, is crucial. They will believe in the fact, that church, is crucial. They will also believe in the fact, that this makes no sense. And, when they go, there, they, might think that this is evil. This is because of the dichotomy of good and evil. This is about the way the evil is about the way the evil is about the way the world is about the content and context that is about the establishment. Now, this makes sense. √

These, are about, the ways, that the followers of Satan, are about the dichotomy of good and evil. Yet, they do not see this, until they become it. This is about the way, that the choice, is about the opportunity and about the direction that it follows that are about the ways that the world is about the direction. This is about the way, the direct, choice, follows. This is from, the dichotomy of good and evil. This is what makes sense.

This is about what the direction, is about, that is about the direction. This is what follows after, the Kings. These are men, in Heaven, I believe, that are about the control, and authority given to them by Jesus Christ. If I were to design the Garden of Eden, this would be true. This would be a God-given position, by the real Jesus Christ. He is about the way the direction, follows, from the truth, from the

choices, that are from the opportunity, that is and are about wealth.

These are the things that follow.

NATURALIZATION- PRESERVING THE ESSENCE-

This is about the way, the manipulators, are a part of the con artists, that are a part of the essence. This is about the essence of hell, and how it does not exist, in this universe. Yet, there are still people, who believe in this. This, is about the way the world is about the way the power is about the unexplained, in the universe, and is about and from the Godhead. This is about anything you believe in. This is also about, the earth, and what this is worth is the greatest substance. These are about the elements, and how they fit into place.

The earth was created. The sky was created. The lesser evil, of the evil world, is naturalization, instead of evolution. The light was divided from the dark. Man made the fire. Man created the knowledge. This is after he was tempted and fell. Jesus created the water. The wind, was always around. And, man created the air. This is through the temptation, for Adam to fall. This is into the deep wide open. These six elements, created what we call, the modern world. There are water, for drinks, there is air, to breathe, there is sky, for ourselves to look up to. There was fire, to be discovered by man. There was the earth, to be created by God. There is outlook to be created by man. And, there is sky, to always look up to. This is about the way, that God created the universe. He divided the dark from the light, which was fire. He created the earth,

through His Spirit. He created, the water, which was from the early beginning. He created the dry land. He created wind, that was always around. He created sky, which was a never-ending frontier. He also created the universe. This is how He did this. He fashioned man, and woman, after each other, and when they fell, they created all of this through knowledge. Yet, they were not in a Garden, so it became survival of the fittest. There are the controls of knowledge, that are controlled by knowledge, that are controlled by knowledge, that are about the world. The world is a perfect fit, for man, and us even after Adam fell. This is still monitored and created by God. This is after man fell. He fell into the deep dark abyss of hell. This is because of his immorality, and condemnation, and superstition, until Jesus came. He performed rituals and practiced, magic, and the arts, and sciences, of the world. This is until Jesus Christ came.

The naturalization form is about the way the naturalization is about the naturalization that is about the naturalization. These are about naturalization, that are about the naturalization, that is about the naturalization, that is about the naturalization, that is about the naturalization, that is about the naturalization.

This is about man, and beast, battling it out. This is about the book of Revelations that are about the way the world, looks, to my man. This is about the dichotomy of good and evil. These are how we fell.

The naturalization, of this, explains that elements that make up man's knowledge are about the equation. This is about the equation, which is about naturalization. Deformation, is the art form of this, and is also the explanation science, that becomes with you. This, is truly about, the

way, man's directional knowledge, are about the direction, of knowledge, that are about the useful manners, that are about our approach, and God. This is about God, that is about God that is about God.

These elements, are about the direction, of our outlook, through the perception.

These variables, of the naturalization, are the variables of our perspective.

Together, we get to see the whole full of knowledge.

These are about, the ways, that the world are about, the ways that the world are about the ways, that the creation of the world, was not this. This is because the creation, was before knowledge. If naturalization, was true, then you would think that it must have made the fall of man, happen, to the universe. Yet, knowledge shows that naturalization is about the certain and inevitable ways that are about the ways, that are about the ways, that are about the ways, that are about the ways, that are about chimps and Valedictorians. These are the ways that the world gets along. These are about, the ways, that the world, is about the ways, that the world, is about the ways, that the world, is about the ways, that the world is about the, God of the world, and about His choice. He is about the way naturalization is about the way that God, is about the choice, that is about the decision, that is about the direction, that is about the opportunities, that are about the outlook. This is of the whole of man. Naturalization explains how the world, is about being united, and that the world must have some order, or some way to explain it. This is about Christ. This is about the opportunity that is about the cost. There are costs that are about opportuni-

ties that are about opportunities. This is a never-ending cycle.√

These are the cycles of just cause and relevance. This is about the way, that conflict and choice, are about the co-ordinates. These are of naturalization. These mean we have free choice, and free will. These are about the myths and tragedies, of our modern era. This is about the choice, for freedom, or the choice of conformity. These are about, the choices, that are about, the opportunity. These are all in naturalization. These are about the ways that the entire world, does naturalization, and about how we know naturalization, to be true. This is about naturalization, that is about the conformity. This is also about the essence. This is about the way the world is about the way the world is about the way the world is about the way the world is about the way the world is about the way the world is about the essences and substances, that make up the apprenticeship of justice.

These are five people, who I think would make good lawyers.

These are, 1. Andrew Dyer, 2. Robert Rowling Jr., 3. Stephen Hawking, 4. Marshall Smith, and Satan. These are about how these are good with the law. They could become great ones. They are the traitors, to the evil and hate of the world, and decide on things of this nature and matter. This is because they are beautiful, creations of God, also. This is according to the way they see the law, with girls especially.

I think the government will frown upon this, when they see them. This is because they are Republicans, who are young and white, and are the great one, known as Stephen

Hawking. These are about and because of the way the world is about them. They are more attractive, than me, which is why I wrote down that they are beautiful. This is from a girl's perspective. I hope this girl, likes me, as much as I like them. They are my only competition. This is because of their 1. Incredible social skills, and 2. Billions, 3. Intellectual prowess, 4. School leadership, and 5. Fallen angel. These are the ones who are about their looks. They see into the artificial outlook of Highland Park, and do try to conform to this, which is about this whole scheme of things. That I just talked about. These are the societal elements, that make up naturalization. This is about the way people think. This is the example of society, that is about them, and their looks. This is about their example, to me, on my computer, that is about the way, that my image, is about Highland Park, that is about the motto. This is always changing, and not ever, the same. This is a lie, a cheat, and a steal. This is about how they conform to Highland Park. This is about their ways, that they look at Highland Park, and frown. Yet, this is a smile. It is from me. This is about my favorite people. They are my favorite people, because of the way they act, and the way they look. This is about a superiority complex I get when I am around people, that is a social disease, of the mind. This is called "popularity". According to no one, popularity is a curse. This is why. Highland Park is not the curse, but the "popularity" of Highland Park. This is about false Gods, false images, false impressions, false acting, and false impressions. This is about the ways that Highland Park, also eats this up. These are what the images of Highland Park are about and what these images are about is about the popular group. They control everything in Highland Park.

How Highland Park is a Mock Scale of the Whole World-

These are about the graven images that are about the conclusions. These are about the whole worlds, that are about the mockery, and shame, that is about the whole Highland Park. Some think that it is hypocrisy but it is actually like any other place. This is a mini world. This is about the ways, that the mini worlds are about the graven images. This is about selling your soul to Satan to become powerful, awesome, and more impressive than anyone else. This is about the ways ,that the images, of Highland Park, are about the ways that Highland Park is about the way the high school, is about graven images too, but not that much. This is about the way Kings, are about, the way the society uses them, to complete the project. This does not make sense. President Kings are for the people.

President Kings are the people. These are the people, in their graven minds. They cannot get out of this. This is due to Highland Park's popularity. This is actually about the world, and how the real world, is about the graven images, that are about the popularity, and the forgiveness, for not letting me succeed. This is actually, about insanity. When, you sit, at home, and watch TV all day, then you start to realize that the television is about you, and you alone, and you get all the attention from this. This is how the world is compared to Highland Park, by me. These are images of mankind that are about the images. This is what naturalization is all about, that is about what it is really about. This is about the way the world, is about the graven images, that are about Highland Park, and are about the frame of mind. This is about the ways, that the acts of images, are about images. Why, can we not see, the hidden truth, that is about our world, and about our graven images. These are about the, graven images, that are

about the ones, that are about the real ones, that are about the completion of the project, and about our own personal satisfaction, that is designed, from the free idea, that becomes with us, and not us, in full. This is about the way, the world, is about the completion, of the project, that is about the ways we learn, and do business, and play. This is about the mind, and how it does reflections, that are about the ways, that we succeed, and are about the images of the mind, that are about the excelling of man. This is because we all live in God's image, and this is the image, that He created the world with. These, are the images, which form and conform to us. These do not conform with us, unless we are a genius. These, are, about, the aspects that are about the personal aspects that are about the client and the student. We "buy" into the world, of images, that are about us, when they are really everyone else's. This, are, the images that are about and from the client and student, and the way that they make us see. This is through the eyes of God's image.

These are about the ways that the guidance and strength of the city, is about the mayor that is about the Senator, which is about the American tradition that is about the ways, that we have conclusions, and that we conclude, to this order of thinking. These are in terms of the progress, which is about America that is about Great Britain, which is about the conformation. This is about the client and student, and about how they, conform, to the entire world. These are of the great one.

These, are, about thoughts, which are about the beautiful relationship, between man and woman. This is about the seclusion, of society, and about how, we conform, to the society. This is about the graven image. These is what the images, is about, and are what the images, are about, are

about the concise, images, that are about the ways, that people, conform images, in their minds, and then the conclusion, is about the way. This is about the way we have 1. Outlook, 2. Conformity and 3. Greatness that follows us, in the conclusion. These, are the only answers to the conclusion. This, is about, the conformity that is about the agenda, the schedule, and the matrix, that we conclude with. These are about and with, the issues, which make us, happen, and have us become. This is about bills in Congress, and how we know about these, is through superstition. These are what control, reality. These are merely, a matter of opinion, that is about these relationships, that is about the intelligent design, that is about the hunger, of the world, and about how it is hungry, and how about power, that eats it up, and a positive message that ends, nothing. These are about the outlooks that are about the conformity that are about the greatness that are about the conclusions, that are about the great ones. These, are about, these conclusions, and about how, we all know what they are. This is all but Perdipa. This is why I want him to become a king. This is about, the client, and the relationship, that he has, with people, of other origin, color, and status, religiously and socially, that pay respect to his country, and about how he leaves his country, to go to New York. This is to hell. This is until he adapts, to the rules, and standards of the club. These are about New Yorkers and how they would have eaten him up, if it weren't for me. These, are about the students of Young Life camp, and how I was his client, so to speak, in generalizations. These are about how, Christ, that knows all, is about the relationship, that knows all, that is about the opinion. This was that I was crazy. This is for Christ. This is how Perdipa found me. This was, all alone, looking for a friend, that I was in Him. This is about religion. This is

about the lies. This is about the deceit. This is about what my whole life was, until I met Perdipa. This was because he was not Satan. Yet, if one man's opinion, is that I am unpopular, this means that I am unpopular. This is the standard of society. This is to be part of the music. This is to be part of the games. This is to be part of the conclusion. This is what I saw, how I saw it, and what I became. This is a true Young Life Christian. This is about what, I went through. This is how I went through it. This is what I became, finally, when this happened. This is about the popularity, of the world, and how it was Perdipa. This is the only friend, I have ever had. This is how popularity does this to you. This is unless it stabs you in the back, and leaves. This is unless it is Christ's. I have learned much worse in school, after Young Life. Yet, I seem to do much better, and be much smarter. This is through slowly learning as a Christian. I hope New York, accepts Perdipa back. This is what I would like, and would want, from the school, I went to. This is what I would like from the innocent games, that we play. This is what I would like, from things that I would like to have. This is from an eccentric mind that I have after writing 58 books. This is where I would be without Perdipa? Yes. But, in the wrong direction.√

This is about, the copycat, con artist, known as George Bush's hater. This is also society. This is about the rock against Bush, and the "anti-W" stickers, and the protests. This is what this man singlehandedly went through. He is a God.

I WOULD LIKE HIM, TO BE MORE LIKE A CHRIST.√

This is about the government, and how these things, go through things, that are not of the opinion, of Christ. These are about haters in the White House.

I hope Henry Hager, is not one. This is about the Bush family, and how they are the Supreme Beings, of the world.

This is the skull and crossbones club. I think the leader of this should be Barbara Pierce Bush. This is the real George Bush's daughter. I think my books would impress her. This is about how she is an intellectual, because she is from Yale. This is what I would like, to marry her for. This is her personality. I just liked it when I looked her up, on the Internet. The way she looks is incredible. This is about her personality, and how the way she moves, makes sense to me. She is also in the Bush family. This is about her, life that is about the public image. This is about, why, I want to marry her. She is in the Bush family, and the Bush family, is the most powerful family. This is also about the Miller's. Which way, that the real Millers, are about the real Millers, is about the Kay Miller family also. These, people, I have a strong bond, with. This is how the world, is about how the world, is about how the world, is about personality, that seems to be what the life, is about. I hope, that I am as cool with the Miller family, as what I am as cool with the Bush family, is. This is family.

This is why I want to be in this family.√

I want to graduate from Oxford University. This means, I would like a Bush scholarship, as well as his friendship. This is because I am a King. I am a power. I am the book of naturalization. I am the body of Christ. This is not true. Yet, I am, a conservative, Christian, who likes to be on top. This is of the game. This is how this is about the game. This is about the conservative Christian, who was a DCC member. This is about how I did this. I did everything with my friends.

I want to be King of the world, with my personal assistant, George W. Bush.√

I am a Godly man, and do not like evil. This is what my rules and laws will be against. These will not be plagiarized or stolen.√ I am not a Satanic person. I have been saved by Young Life. Yet, this is what troubles me. This is that the skull and crossbones are a weird organization.√ I want to be King, of every nation, and country. This is what my position will be. This is of King, or President King. This is about the way that God moves in people. They are the Christians, who are sealed in the Book of Life. I do not write in the Book of Life, and do not condemn people. This is impossible for me to do. This means that I am a Christian who has been in trouble. This is only with my father. He has put me through everything, once, and I still do not talk about it. This is only with my friend, Justin Fanning. He is a good influence on me. I think he should also work with me. I think that God is about people who are about the good in them. This is about how I have never written in the Book of Life, and do not think that anyone can do this, but weird people. These are about trials and tribulations that are about these kind of people. These are in every civilization and every world. These are the people proclaiming that they are God. This

is because of what I have learned about it. This is that it contains annoying and contagious, false God, syndrome. This is about the way the prophets and the saints are not part of these. These are the everyday good people, around the world. This is about the way the beings around the world, are. These are the evil people, who think they can judge people, and interpret my book literally. This is about the way that people are about the constitution and how it also cannot be rewritten. There will always be history and it will always be interpreted correctly. These are the showoffs of the world, and what they think, does hurt me. This is about my compassion agenda that I have as a future trillionaire. This is if they accept me. These are about the councils and wisdom of the evil. They are about money, lies, and stealing only. These are not how I plan on writing in the Book of Life. This is about the way the Book of Life is about the written constitution and how the constitution is about this is about how we all follow the law. This is of God's. This is about how God is about love, and compassion, and not evil. He is about the trials and tribulations of mine. This is but when I did something I shouldn't have done. This is not to be talked about. Yet, I feel a strange feeling that Satan is still around. He is the one who made me do this. This was something with a girl once. It was not sex, and was caused by her. This was nothing of the way you think it to be. This was just me for three minutes, kissing. I think that this is wrong, because of Satan. The only persona; savior who I want to kiss, is my wife. I do not want to sin, or do things wrong, anymore. This is because of the real John Miller. He is my best friend. Yet, he is the worst influence. This is also because of Satan. This is how Satan works. He is not part of the church. He is not the skull and crossbones. He is an angel that walks the earth. He

can become other people. He can also become strange with others. He is the one who practices magic. He is the one who strangely becomes with others. He is part of the human race. Yet, this is as an angel. He is. He was. And, he shall be. This is until his afterlife. He is part of the human race, because he is intelligent. This is the way, that he deceives. He even deceives God. He is a lunatic, also, on purpose. He is the angel that fell from Heaven. He is the one who goes to the church, in disguise. Yet, he cannot do harm, to others. This is the devil. This is what the devil does. He hunches over, and types on my computer. Then, he sends me to jail. This is what is wrong. He is inhumane. This is how I interpret Revelations. This is through the Bible. He is also in the Bible. He is about the ways the Bible, is about him. This is about the way that he does things. He is about the person who is about the way who is about the deformation. He tricks people, into doing deformation, the wrong way. This is about the whole of my thoughts, and are about the way the world is about the way the world is about the way he is not anyone. He is just evil. This is about the way that the evil, of the world, is about the savior Jesus Christ. He is what Young Life calls God. He is about the way that they are not evil. They are just participating. This is in the world. If you want to be saved, it is personal. This is about the way people are. They are not betrayers or liars they are real. They are about the way the world is about the way evil controls it. This is very true. The evil of the world is about the angel Satan. He is not in my family. He is not a Book of Life person. He is the way that I do not become King. This is unless Bush endorses me. Thank you. I plan on starting the new world order and carrying it out. I am not a agent of Satan, or the Devil. I am not a participant in evil. I am a good person. This is not about how strange I am. This is

about how smart I am. I am very smart. This is about the way friends, are also this way. They are the ones who admire you. They are the ones who bring you up. They are not bad. They are not part of the skull and crossbones, because this is a secret society. They are about the way that the world is about the way the world is about God. He is about the way that society is about the way that society is about God. He is about society. He is anything you want Him to be. He is always capitalized. He is also always good. He is about the way you believe in Him. He is about the way you know Him. He is about the way we know Him. He is the honor roll student. He is the leader. Yet, He is anything you want Him to be. He is the snake, in the family. This is about the way the doctor I went to, was a devil. This is about the way he falsely diagnosed me, with a disorder. This is true according to my honor. He is the one who diagnosed me, with a false diagnosis. He is Dr. Kenneth Phillips. He is very bad. He was the one who tortured me. If I am a person, I do not deserve this. I deserve to be treated fairly. This is not what John Miller thinks. He is the person, who beat me up some. He is also the one who lied to send me to Dr. Phillips. This is true. This is about how the law, is about me, that is about God. This is about God, and me, Riley Parker Miller. This is about the logic, of me, Mister Riley Parker Miller. He is a God, and makes sense to man. This is about Dr. Phillips and how he makes sense to me. He is about the snake, and the mongoose, and diagnosing this. This is about Dr. Phillips and how he died. This is about Jan Miller, and how she didn't die. This is about Dr. Phillips and how he died. He is about a God, and how He died. He is about Dr. Phillips and how he died. This is about Dr. Phillips and how he died. He died from cancer. Dr. Phillips dies to cancer. He is a doctor. He is dedicated, to his dying. He

does take care of dying people. This is in a good way. This is because he died, to died. He died, to his own dying. He dies. He is about dying, which is about dying. He is about dying, that is about dying. He is about death, and is about death. He is about death, which is about death. He is about dying, that is about dying. He is about death, that is about death. He is about dying, that is about death, that is about the death. This is about dying, that is about death. He is about, dying. He is obsessed with dying. This is because, he is death. He dies, a death every day. This is because, he has patients. He has patients, that have patients. This is about death, that is about death. This is death.

This is what he does to people. He diagnoses the dead. This is without Jesus Christ. This is about diagnosing dying. He is about dying, that are about dying. Death is all about the death. Living is all about living. Dying, is about dying, a death. This is all about death, that are about death. The death, of death, is about the death. This is about dying, that is about dying.

This is about death, and death. This is about dying, that are about dying. Naturalization is all about naturalization, that are about naturalization, that are about naturalization, that are about dying.

Naturalization is about death that is about death. It is about dying, to sciences, and then living. Then, it is about, another life. This life is about another life, that isn't about life. This is about death. This is about death, which is about death. This is about death, which is about death. This is about death, that are, about death. This is all about death, and about how death, is life. This is life eternal. This is about death, which is about death. These are about

deaths, which are about deaths. This is about deaths, which are about death. These are about death. This is about dying, that is about dying. This is about deaths, that are about death. This is all about, naturalization.

This is about life, and life. This is about what is about, the death. This is all about life. This is about natural things, that are about Jesus Christ. This is about how life, is about death. This is about how, death is about life. This is all about, naturalization, that is about naturalization. This is about how God, walks the earth, and is about Him in Heaven. He is about the Satan. God is about, the Satan. He is, about how, He is about how He is about, how He is about, how He is about. God, is about the Satan. Satan, was with God, in the beginning of time. Satan, was a Serpent, and fell from Heaven. This is according, to naturalization truth. Heaven, is about all angels, and about all love and works. God, is about Satan, and superstition. God, is about Satan's fall, who was an angel. Satan, fell from Heaven, without God, and without Jesus Christ. He is about an angel, who fell, and who fell forever. God is about, an angel, who is about an angel. God, is about an angel, who is about an angel. God, is about Satan, before time. Satan, is about manipulating time. Satan, fell from Heaven. Satan, is a false, angel. God, is about judging, a false angel. God, is about manipulating, time and space. This is, because of an angel, who fell from Heaven, who tricks us, and manipulates us all. He is the chosen one. Satan fell. Jesus Christ, did not fall. Satan, who fell, fell from Heaven. Heaven, is a place, driven by success and determination. Jesus Christ, did not fall from Heaven. Satan, is an angel, who deceives. He is not around anymore. This is because he is like Doctor Phillips. Yet, Satan hates, instead of heals. Jesus Christ, is the one who de-

feats Him. Jesus Christ, is all of the world, that is all of the world. God, is about Satan, who is about Satan. Satan, is a fallen angel. Satan, is the fallen, angel of the Lord. This is about Heaven, and earth. Heaven, is ruled by Jesus, and earth, is ruled by Satan. God, is about an angel, of the Lord. This is about earth, that is about earth. This is about earth, that is about earth. This is about how, the earth, is ruled. The earth, is ruled by an army of darkness, that is controlled by Satan. Church, is about what, defeats, the army of darkness. None, can boast of the wisdom, and brag. This is unless, you are Jesus Christ. This is about, the God, of the naturalization. This is about naturalization, that is about naturalization. This is about how Satan, is defeated by Satan. This is all about, naturalization, and about how church and state are separated, by a void. This is of Satan. God is, the real one, who is in control, of all, of the earth. And, Satan's army, is defeated. This is by police.

This is by Naturalization, and All About How Christ is Defeated-

This is about the password system. This is about how we pray, and how we think. We think of all good things, all the time. I wrote "Accolades" which is a book, by me. This is a book, by me, and by the antichrist. He is the one, who solves the world's problems. He is the "trickster God, known as the spider". God is in control of all of this. God is in control of the world. God is immense, in knowledge, and the immaculate conception that He controls. God, is about the power, and about the real authority, of Young Life, adventures, in a camp that made no sense. This is because some, are traitors, to the government. This is because they are Christ.

"There are no such things, as the real "Antichrist".

This is about a real fiction. This is about Young Life. They are about defeating, Satan. They always defeat Satan. They defeat Satan, in everything they do. So, does the world. And, so does He.

He goes to YL, camp, sometimes, and prayed to Jesus. Jesus is the one who prays to the world. Young Life camp is the most impressive camp, in the entire world. The password systems, is what they want. This is about what they get. They, get candy, at different camper's tents. They, also, get chocolate, at different stations. This is, but, to solve Satan. They must, defeat Him, in all that they do. This is because, that they have, a big ego. I think Satan will be defeated in the end. This is about what everyone will know about me. This is that you do not mess with me. This is through the police, to the firefighters. This is from the world, to the underworld. I think that Satan, loses because he is part of the underworld. This is about, the underworld, and where is the world. The worlds, are part of the world, which is part of the world, that is part of the underworld.

Satan is part, of the world, that is part of the world. He is a part of, the world, and the underworld, that is a part of the underworld. He is a part, of the entire world, that is a part of the world. This is about Satanic, which is about the real Satan. All is about the hidden Satan, which is about Satan. He is the great deceiver, which is the great deceiver. He is about the Satan, which is about the Satan. He is about the hell, which is about the hell. He is a part of the underworld, which is a part of the underworld. Each people, and people, have a personal key, which can and will and could defeat Satan. He is merely, a, in charge, and

immediate angel, in charge of the underworld. Christ, also defeats Satan.

By Jesus Christ, He can and will, defeat satanic cult. Jesus Christ, can defeat the intelligence driven Satan, by being the real Jesus Christ. He is, through the eyes of the reality driven illusions, are a part, of which is about the underworld. Satan cannot be defeated by Jesus Christ, alone. It is Jesus Christ, that is about Jesus Christ alone, that is about the illusive Jesus Christ, alone. Jesus Christ can and will, defeat the real Satan. But, there is a real direct, angel, that has a key, to the endless abyss.

Satan is also always, about the underworlds, which are from the underworlds. Jesus Christ, always defeats Satan. Jesus Christ has to defeat the real one who is the chosen Satan, of the world, known as the Biblical and personal Satan, to the churches, of the world. He is about Satan, which is always, about Jesus. Jesus Christ is about the underworlds that are about satanic people, and the church, of the body of Jesus Christ, which isn't Satanic. The real and true Satanic people, are about, being the always and surefire churches of the real defeated one, known as the real world, of satanic churches, that is. Satan seems to be about the losses, of satanic people, and the real satanic church.

Christ is about winning, at the game, that is life. Satan is for death, of the illusions, and the underworld.

These are about the losses, of about the losses, in which are losses. The real body of Christ, that defeats Satan, defeats the real one, by believing in the church of Christianity. Satan, is all about Satanic and unbelieving in the church, of Christ. The satanic bodies, are all defeated, by the real Jesus Christ, that are in the real Jesus Christ, the

body of Christ, and the church of Christians, of the all the church body. Jesus Christ, in the ending, will defeat Satan, by marrying the church, known as the real bride. The Spirit of the churches, are the real Jesus Christ's alone body, and is about direct, impressions of and the results, of the hidden agenda, that is of Hinsdale, that are about illusions, and that are about hell, that is about earth. And, the groom is the real, Jesus Christ, and the real bridegroom. The real church, are about the bodies of Christians that are about the churches. God is about satanic realities, of pain and hurting, that are about Heaven, and Heaven alone, that is from Satan. He is the biggest, con artist, and scammer in the history of the real world. That, is about and of, the real God's.√

The Lord God Almighty, that is in all of the charge, of the church body, of the real Christians, is in charge, of the church body, known as Christians in the world. This is a part, of the real, Jesus Christ. This is about the real Jesus Christ, that is in charge, of the real church body, that is in charge, of the Christ's body. These are, about the real people, that are in Christ. The body, in the church body, is in Christianity. This is of that all that is a part of the real people. These are in charge, of the real person, of the real Heaven. This is about, what is always about, what is also Jesus Christ. Jesus Christ, is in the all-knowing knowledge of God, that will defeat the real Christian church. This will defeat, the church, into all knowing.

This is a King Written Piece solely by the work of Riley Miller.

Mr. Riley Parker Miller

5800 Royal Lane Condominiums- 10711 Villager Road; Unit C.

Dallas, Texas 75230.

RPMDallas1@gmail.com

Preston Hollow

www.ingramcontent.com/pod-product-compliance
Lightning Source LLC
Chambersburg PA
CBHW021810170526
45157CB00007B/2528